Tripping Over Engineering: Going Nuclear

Exploring the Birth of the Atomic Age

by

Lindsey Bredemeyer

Copyright © 2018 by Lindsey Bredemeyer
All Rights Reserved.

ISBN: 9781790174973   Paperback
ASIN: B07LC1Q6RH   Kindle EBook

Title - Tripping Over Engineering: Going Nuclear
Subtitle - Exploring the Birth of the Atomic Age
Author - Lindsey Bredemeyer
Publisher - Lindsey Bredemeyer
Series - Tripping Over Engineering - 1st

Contact: TOE@LBredemeyer.com

CONTENTS

Personal Introduction 6
Background on the Book 7
How to Use this Book 8

NUCLEAR MINDSET 9

CREATION 10
Chicago Pile-1 (CP-1) 11
X-10 Reactor 14
Enrichment 18
Hanford B Reactor 21
Fireset 27
Trinity 29
Fermiac 32
Redstone Arsenal 34
Nuclear Testing 37
EBR-1 40

DELIVERY 43

BOMBS 44
Mk-1 Little Boy 45
Mk-3 Fat Man 48
Fat Man Derivatives 50
Mk-7 52
As Big As You Can Carry 54
How Much Is Too Much 56
Palomares Casings 58

BOMBERS 61

AIR FORCE BOMBERS 64
B-29 Superfortress family 65
WWII 66
Post War 70
B-50 72
B-29 Family Nuclear Tests 76
Derivatives 77

Tripping Over Engineering: Going Nuclear

B-36 Peacemaker 80
B-47 Stratojet 84
B-52 Stratofortress 88
F-84 Thunderjet 92
B-58 Hustler 94
XB-70 Valkyrie 97

NAVY BOMBERS 99

B-25 Mitchell 100
P2-V Neptune 103
   Truculent Turtle *104*
   Carrier operations *107*
AJ Savage 109
AD-4B Skyraider 111
A3D Skywarrior 113
A-5 Vigilante 115

CRUISE MISSILES 118

TDR-1 119
JB-2 Loon 121
MGM-1 Matador 123
SSM-N-8 Regulus I 125
MGM-13 Mace 128
AGM-28 Hound Dog 130
SM-62 Snark 133

BALLISTIC MISSILES 135

ARMY BALLISTIC MISSILES 136

MGR-1 Honest John Missile 137
MGM-5 Corporal Missile 140
PGM-11 Redstone Missile 142

AIR FORCE BALLISTIC MISSILES 145

SM-65 Atlas Missile 147
LGM-25C Titan II Complex 150
LGM-30 Minuteman 155
   LGM-30 Minuteman II Launch Control Center & Silo *157*
   LGM-30 Minuteman II Training Silo *159*

SURFACE-TO-AIR MISSILES 161

Tripping Over Engineering: Going Nuclear

MIM-14 Nike-Hercules	162
CIM-10 BOMARC	165

AIR-TO-AIR MISSILES 167

AIR-2 Genie	168
F-89 Scorpion	170
AIM-26A Nuclear Falcon	173

BIG GUNS 175

M65 Atomic Annie	176
M-28/29 Davy Crockett	179
W23 Katie	181
W33/W79 8 inch	183
W48/W82 155mm	185

POWER 187

USS Nautilus (SSN-571)	188
USS Enterprise (CVN-65)	190
NS Savannah	193
HTRE-2/3	195
Supersonic Low Altitude Missile (SLAM)	198
NERVA	200
Phoebus 1B Nuclear Rocket	*202*
Nerva XE" Nuclear Rocket (AKA XE Prime)	*203*

GLOSSARY 205

[LOCATION BY STATE] 209

[LOCATION BY SITE] 210

{SOURCES} 217

Tripping Over Engineering: Going Nuclear

Personal Introduction

I was born the week after the Cuban Missile Crisis, five miles from an active Atlas ICBM site, 30 miles from a B-47 base on nuclear alert, and the local antiaircraft missiles even had nuclear warheads. By the time I started school, most had given up "Duck and Cover" training, but mine continued since nearby Dyess AFB maintained B-52s on nuclear standby alert. Four years before I was born, Dyess had a Broken Arrow incident with a B-47 crash on takeoff and a nonnuclear detonation of the ordnance. While on active duty 1984-1988, the Coast Guard did shipboard training based on response to Soviet nuclear attack.

In engineering we strive for objectivity in the product of our work, but we often pursue our work for subjective enjoyment. I know perfectly competent engineers that punch the clock for the company and have entirely non-engineering lives when they step out of the office. I never understood the guys that didn't see how cool their work actually was. As far back as I can remember, I have had a fascination of science, machinery and technology. So even when I travel, I am looking for hit of sweet sweet engineering. There is a lot of cool stuff out there, which in many cases is publicly accessible. While I appreciate you reading this book, the true measure of success is your own experience of these sites. Bon Voyage.

Background on the Book

The progress in nuclear physics through the 1930s provided the theoretical understanding that made the development of nuclear weapons inevitable. Someone at some time would have eventually made the investment to proceed. It is more of a recognition at what can be achieved when unlimited resources and properly focused management are thrown at a particular issue. And this wasn't the development of an improved ship or aircraft, it was a device with a function that was entirely theoretical only three years before. The Wright brothers already knew that kites and birds flew, they needed to mechanize the process.

At the time of the first nuclear detonation, Germany, Japan and the USSR were all pursuing some level of nuclear weapon research. The nuclear cat was already out of the bag, it required a lot of money and effort to train that cat for your own purposes. During WWII, Germany had made compelling advancement so that specific missions were tasked to destroy facilities supporting their atomic projects.

Once the war was over, there was a brief hesitation in nuclear weapons development, but the growing Soviet threat reinvigorated work. The expense was always justified with a crisis, which in retrospect rarely equaled expectation. A bomber gap, a missile gap, a battleship gap? Into the 1970s, the Soviet Kirov class missile cruisers led the US to reinstate the Iowa class battleships one more time, bet you didn't see that one coming. Dr. Strangelove even addressed the Doomsday Gap {Kubrick 1964}. But, the Soviets were not the only fight over nuclear weapons. The Army, Navy and newly minted Air Force of 1947 were all competing for the diminishing post war military budgets by expounding the necessity and superiority of their technology over the others. Every group thinks their ideas are superior. The more so when the technology is still in diapers.

How to Use this Book

This is not an exhaustive catalog of the subject. It started with where I have been, but I found that there were some artifacts that deserved mention. There are also those that may be inaccessible, but are deserving of recognition. There is always more to be added, but a finitude of available time. There is always the potential of a next book.

Categorization is a double edged sword, both providing structure while restricting organization. I have attempted to provide a reasonable sequence of progress.

Locations: All locations are indicated in [] referencing the specific [Location by Site]. Sites by geographic location are shown in [Location by State].

Sources: All sources are indicated by {} referencing the {Sources} section.

NUCLEAR MINDSET

In 1945, the US had six bombs and 15 bombers capable of delivering a total of 120 KtE yield. By 1959, there were 2,500 warheads and 1,500 delivery systems. The peak inventoried yield occurred in 1960 with 19,000 MtE total. The peak inventoried warheads occurred in 1987 with 14,000 total warheads for all delivery systems {Miller 2001}. This should convey a very good concept of MAD (Mutually Assured Destruction). The repercussion of attack is annihilation.

The Cuban Missile Crisis in the Fall of 1962 is always the gold standard reference as the brink of armageddon: {Dobbs 2008}

Sat 20 Oct, there was a cabinet discussion of whether to blockade Cuba or make a full on air strike. There were many in favor of the air strike. The blockade finally won out for the simple reason that, the missiles may have been fired purely as a retaliation to the air attack. Knowing full well that they would have retaliated if the Soviets had wiped out US Jupiter missiles already armed in Turkey.

Mon 22 Oct, Tactical weapons were now extensively in the fleet. It was decided to provide various single seat fighters with nuclear weapons and disperse the aircraft to a broad range of airfields to reduce loss from a nuclear strike. This also functionally bypassed the two man rule and placed the weapons outside of the typically accepted secure custody procedures.

Tue 23 Oct, Second Soviet ship arrived in Cuba, bringing the Soviet total to 158 warheads with 37.2 MtE for bombers, cruise and ballistic missiles.

There was a US population of 92 million within range of Cuban missiles. Civil Defense had extensively promoted the establishment of fallout shelters. A total shelter space was available for 640,000 of which 170,000 contained rations for anything over a days occupation.

Sat 27 Oct, 60 B-52s are on airborne alert at all times, 183 B-47s on dispersed field standby and 136 launch ready ICBMs.

Sun 28 Oct, By mid day there was a ready to fire US nuclear force of 162 missiles and 1,200 planes totaling 2,858 warheads.

Tripping Over Engineering: Going Nuclear

CREATION

"It is still an unending source of surprise for me to see how a few scribbles on a blackboard or on a sheet of paper could change the course of human affairs." Stanislaw Ulam {Rhodes 1986}.

Tripping Over Engineering: Going Nuclear

Chicago Pile-1 (CP-1)
Manhattan Project
University of Chicago
2 Dec 1942
1st controlled nuclear fission

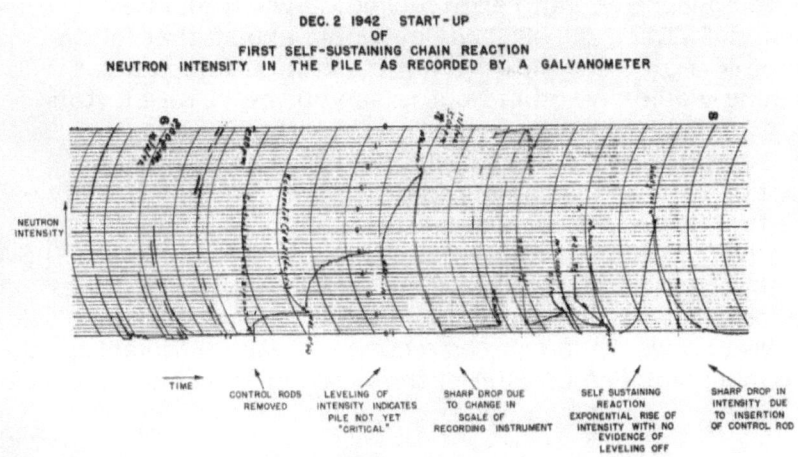

Birth certificate of the atomic age {DOE}.

With most of the Manhattan project, nuclear facilities were placed in remote locations for both security and safety reasons. Logic would follow that the first sustained controlled nuclear reaction would have the same criteria. Or, just build it in the middle of the second largest city in the US. In a squash court under the Stagg Field west stands at the University of Chicago, Enrico Fermi led a project to build a pile of 771,000 lbs graphite, 80,490 lbs uranium oxide, 12,400 lbs uranium metal {Rhodes 1986}. Hence the name, Chicago Pile-1.

CP-1 was uncooled and unshielded. There was a set of control rods managing the fission using wood strips nailed over with cadmium sheet {Rhodes 1986}. There was also a safety control rod with an elegant control and operation system. Norman Hilberry Ph.D. had an axe with orders to cut the rope supporting the weighted safety rod if something

Tripping Over Engineering: Going Nuclear

should go wrong, a well specified criteria. There was also a last ditch safety system of a team with buckets of cadmium solution. If all other safety systems failed, they were to run in and dump the buckets on top of the reactor. The solution would seep in and permanently kill the reactor. Fermi still had to convey the safety of the project. It is reported that when asked by the university president what he would do if anything went wrong, Fermi replied, "I will walk away-leisurely" {Rhodes 1986}. One would expect that in the event of a problem with the world's first nuclear reactor, running would do nothing but raise your heart rate further.

On 2 December 1942 the reactor achieved a sustained reaction with an instrumented growth of neutron intensity. The first sustained nuclear reaction ran for 4.5 minutes at a massive power output of a half watt, yes that is correct, 0.5W. Or about the same as what is defined as a USB low power device. It was eventually operated at 200W, which my generation would refer to as a couple of standard lightbulbs. The most significant result of the reaction was the breeding of Pu239.

CP-1 operated at the University of Chicago until 28 February 1943. It was then moved, reconfigured, shielded and placed into service as Chicago Pile-2. CP-2 operated until 15 May 1954. Parts were reused on other tests, the remainder encased in concrete and buried at Disposal Site A. A marker stands at the reactor burial location in the Red Gate Woods near Chicago. A plaque was placed in 1952 on the Stagg Field West Stands, which are now gone. Nuclear Energy, a Henry Moore sculpture, stands at the location. Argonne National Laboratory continues operation nearby with Fermi Lab farther west.

Why Stagg Field? It wasn't wartime austerity. From 1892 to 1939, the University of Chicago was a heavy hitter in college football. In the late 1930's, the university president believed that football was not particularly useful for academic pursuits. With significant public support, the football program was canceled in 1939, not returning to college competition until 1969.

While there are no facilities to be seen, the significance of the

Tripping Over Engineering: Going Nuclear

achievement and unexpectedness of the location make it compelling. Particularly in light of the remoteness of so many Manhattan Project facilities. The progress of the Manhattan Project was extremely dependent on these accomplishments.

A few pieces of the original CP-1 survived. Some of the graphite moderator bricks were retained. While generically refer to as graphite, it was the first of "nuclear grade" graphite. Early tests were restricted by trace boron absorbing neutrons. A graphite with far fewer contaminants was produced using petcoke. Pieces of the first nuclear reactor are at [NATM], [BSM], [MSI] & [NMNSH].

Tripping Over Engineering: Going Nuclear

X-10 Reactor
Manhattan Project & Cold War
Oak Ridge National Lab [AMSE]
4 November 1943 - 1963
1st continuous operating nuclear reactor

Logan Emlett and first atomic electricity generated at X-10 {DOE}.

Once CP-1 verified the ability to achieve controlled fission and produce Pu239, it was necessary to escalate the production rate in order to achieve the required design mass within a reasonable time. Since CP-1 was uncooled and unshielded, any significant power increase would quickly reach temperature limits. It also lacked a convenient production method for the insertion of new isotopes and the removal of irradiated isotopes.

There is a dull gray corrugated sheet metal building with the demeanor of a warehouse storing precious and important

Tripping Over Engineering: Going Nuclear

items such as bales of toilet paper. Upon entry, it is all business. 32 feet high and 38 feet wide. 1,248 ports across the face, with 800 of the channels operating as primary reaction tubes, typically providing around 54 tons of fuel. The remainder configured with instrumentation, blanks or other test isotopes. This permitted testing and adjustment to achieve the best neutron and heat distribution across the reactor. The moderator was still special purity graphite, 640 tons in a 24 foot cube. The required maximum heat dispersion was 4,000 KW, simply solved with three 55,000 CFM fans. Because of the sensitivity of heat transfer to air temperature, the colder the weather, the harder they could run the reactor. The shielding was elegantly addressed by seven feet of high density concrete all the way around. The exchange of fuel rods was simply shoving slugs in the front of the reactor with irradiated slugs dropping out the back into a wet channel for cool down.

Uranium and plutonium have particular corrosion and pyrophoric hazards. To address this, the fuel slugs were aluminum encapsulated for low neutron absorption. But aluminum also has a range of issues with melting temperature and its own flammability. Even the graphite moderator has burning risks once adequate temperature is reached. Should worst case ever occur, you are fanning the flames of multiple flammables with 165,000 CFM. An exhaust filter of 0.00004 inch porosity filtered to a 200 ft stack as a containment effort in the event of an accident. Similar filters in British Windscale reactors prevented their 1957 fire from being a much greater environmental disaster.

In old classic factories, the shop foreman would watch you from a compact elevated wood office with windows omnisciently overlooking their domain. The control room for X-10 is exactly that, overlooking the floor and the face of the reactor. The analog direct reading instrumentation has the aura of being somewhat less aware. There are a range of basic warning lights and gages, a rack of manometers for airflow measurement, and proper rolling chart recorders for a range of data.

Double pushbutton start-stop switches are simply marked insert and withdraw for the two control rods. One would

Tripping Over Engineering: Going Nuclear

think that the mode with the increasing hazard would have a spring return switch so it couldn't just run to max reactivity. Any motion of either control rod continues in that direction until manually hitting the stop button or the rod hits max stroke. These straddle the operator position in the knee bumping location on the front of the control desk. Expecting the start and stop buttons to be differentiated, or at least a consistent control of the two control rods, Rod One insert switches are panel mounted vertically and Rod Two are switchbox mounted horizontally. Not what one would consider an inherently safe design.

A thermopile chart recorder has a piece of red tape at the 4,200 mark and at the bottom of the window on a blue embossed stick on label as would mark Uncle Ed's Rapala collection it says, "SCRAM AT 4200". Above the main control panel is a rack of forty white/red alarm lights. They indicate a range of specific issues with well identified criteria: SCRAM, Hi Radiation, Low water, Low resistivity demin. One warning light that captured my attention, "TROUBLE, HOLE NO.15". As the first continuous operating reactor, what particular measurement would be defined as "trouble" and why is Hole No.15 particularly notable.

Adjacent to the main control panel is an array of supporting panels. There are more chart recorders and an assortment of gages on thermocouple switch panels. Because of the reactor configuration, only specific channels are instrumented, with significant data averaging. A particular level of care was necessary to avoid hot spots. The original control panel was black with lights and white faced instruments which looked more business like than the later soft industrial blue. There are a range of wood panels where mods were made in the panel, but there are also a number of instruments that remained stationary all twenty years.

The reactor went critical on 4 November 1943 and produced plutonium until January 1945. It was then tasked for production of peace time isotopes for uses such as medical and industrial until 4 November 1963, exactly twenty years later. Brookhaven National Laboratory built the first postwar reactor purely for the production of peacetime isotopes. It was an aircooled graphite configuration much like X-10.

Tripping Over Engineering: Going Nuclear

Some things are too cool not to be noted. On 3 September 1948, X-10 became the first nuclear reactor to generate electrical power. One of the tubes was configured with a boiler which was used to drive a Jensen #50 steam engine with generator. Jensen has been producing model and demonstration steam engines and generators since 1932. On this day, while the electrical output is unknown, the engine is capable of a rompin stompin 1/10 hp. It probably lit up a flashlight bulb. For $5000, the #50 is still available from Jensen. But, you have to provide your own reactor. The little engine that could perseveres at the reactor.

Tripping Over Engineering: Going Nuclear

Enrichment

Manhattan Project
Oak Ridge TN [AMSE]
1942 - Varied
Enriched uranium for 1st nuclear weapon

Operators at Y-12 panels {DOE}.

There was a great desire to develop a nuclear weapon with the only naturally existing fissile isotope, U235. The dilemma is that >80% U235 content was necessary for the device and U235 exists as only 0.711% in natural uranium. Regrettably, the separation is from other uranium isotopes of 99.984% U238 and trace U234. Since the isotopes chemical behavior is identical, the only variance for differentiation are the mass and size due to a few neutrons. The only purpose of the Clinton Works was to subdue this process.

Centrifuge:
While conceived in 1919 and now the standard method of industrialized enrichment, the gas centrifuge research for the Manhattan Project was discontinued in 1944 due to extensive mechanical development preventing adequate production in the required time.

Tripping Over Engineering: Going Nuclear

S-50 liquid thermal diffusion:
The S-50 facility used liquid thermal diffusion from Navy testing in Philadelphia. Towers with a hot wall and a cold wall incurred thermal convection in which lighter and heavier uranium hexafluoride would have preferred movement. The basic design is conceptually simple with staged scaling, but the thermal convection and temperature limits restrict the effective energy applicable within a single stage. It was estimated that 100,000 tubes would be required to achieve the specified production with 4,600 stages to achieve 90% enrichment. Certain available power characteristics brought the construction of a 2,142 tube plant.

Unfortunately, the natural convection required significant time to settle into stable production, with estimates up to 600 days to reach steady state. It initially used 1,000 psi steam from the main power plant, but this power was concentrated on K-25 when it was operational. Since this was initially a Navy project, they provided twelve spare 450 psi destroyer escort boilers, significantly reducing the temperature differential and total energy driving the convection. It was found to be unsuitable for significant enrichment, but it did provide preconcentrate feed to the other processes. The most optimistic production claim was that it accelerated the weapon delivery by one week. S-50 is the only known production attempt at liquid thermal diffusion enrichment. It was shut down at the end of the war and demolished before 1950.

Y-12 Electromagnetic:
The Y-12 facility was built for the use of electromagnetic separation. The uranium tetrachloride ions were fired off the emitter with a magnetic field to curve the path of the ions through a vacuum chamber. The different mass isotopes would arc into different traps. Similar to a classic cathode ray tube television. It doesn't have the feel of a mass production device, flinging a few molecules at a time, warped through space with humongous electromagnets. Vacuum containment was often an issue. In order to have enough conductor material during limited times, $1B of silver, 13,300 tons was borrowed from the US treasury. After the final tear down with silver recovery and return in 1970, there was a total silver loss of only 8 lbs. While electromagnetic has limited production for weapon isotope volumes, it has the advantage of a high single pass separation with a low train

Tripping Over Engineering: Going Nuclear

requirement for specialty isotope enrichment. There are portions of a Y-12 Calutron at [BSM] & [AMSE]. [AMSE] has also kept some control panel sections.

K-25 gaseous diffusion:
The K-25 site ran gaseous diffusion enrichment using uranium hexafluoride. This mandated significant development in corrosion protection and chemical tolerance. Since it cascaded stage to stage with only a 1.0043 enrichment factor, the scaling in construction resulted in a U-shaped building with a total length of a mile. The performance was significant to maintain U235 enrichment until 1964.

Originally tested in 1913 using an unglazed clay pipe membrane, Franz Simon wanted higher pressure tolerance with fine pores while preventing excess restriction for greater efficiency. His solution was to take a fine wire mesh kitchen strainer and hammer it to an array of fine pores {Rhodes 1986}. ICI was commissioned to produce the compressed wire membranes. Later gas diffusion facilities benefited from improved membranes and optimized process rates and pressures for the different stages of enrichment. The mechanical pressurization also provided more energy input into separation than the natural convection of S-50. But it was intensely train and stage driven, requiring a lot of material in the system and taking a lot of tiny steps to enrichment.

Gaseous diffusion became the standard enrichment method until the development of the centrifuge. K-25 was one of the most recognized and iconic Oak Ridge facilities, producing until 1985 with demolition completed in 2015. The remaining concrete feels more like an old military airfield than a foundation. [AMSE] has the original Columbia University lab test unit which demonstrated gaseous diffusion enrichment.

The shotgunning of the production methods was necessary to determine which methods were the most productive. It also found in what range of enrichment each method was most effective. This allowed the sequence and mixing of production to allow different methods to feed another step of enrichment even if it didn't achieve a complete enrichment on its own.

Tripping Over Engineering: Going Nuclear

21

Hanford B Reactor
Manhattan Project & Cold War
Richland WA [HS]
Critical 27 September 1944
Operating 1944-1963
1st production nuclear reactor

Hanford reactor complex {DOE}.

In order to free themselves from the tyrannical process for enriching uranium they had to develop the tyrannical process for breeding and extracting plutonium.

Dry and desolate, with a sporadic glimpse of foreboding facilities. A nondescript steel ranch gate with a few federal warnings is the threshold of the original desert. Rattling along a dirt road you eventually amble toward an ominous poured concrete cube of a building. Its demeanor only aggravated by the sparseness of its habitat. Still meeting the memorandum to contractor definition of "a very large isolated site". Hanford B Reactor, the first of three original production reactors which were key to the entire US nuclear weapons program.

These demanded immensely higher fissile rates than the uncooled CP-1 and air cooled X-10 reactors. Like all thermal

Tripping Over Engineering: Going Nuclear

processes, the magic is in heat transfer with one of the best heat transfer mediums, water. The original heat rate for the reactor was 250 MW, eventually escalating to 1,640 MW.

Before entry, the only warning is not to reach where you can't see, because the local fauna either bites or stings. A long windowless institutional corridor of dull artichoke green leads to a mundane double metal shop door, familiar to some of us from old cold war built factories and schools. A roadblock forms as the visitors stop stunned. Working forward, it appears to the left, a mass terminating an expanse, 36 feet high, 36 feet wide, the cold end of Reactor B with 2,004 individual port entries on the face for fuel charging and cooling.

From this end, workers rammed stacks of uranium fuel about the size of a roll of quarters and aluminum spacers into aluminum process tubes. They would have to do this for about 1,500 active process tubes. The remaining tubes were used for instrumentation, testing or isolation to balance the fissile distribution. This furnished a total uranium fuel charge of 64,000 elements totaling 200 tons. As they would drive each new charge into the 1,500 ton stack of graphite moderator, it would push the prior 28 foot long irradiated charge out the back into a wet basin.

Framing the sides of the reactor are two thirty six inch risers. Seventy eight sets of four inch strainers and gate valves managing distribution and isolation of the thirty nine four inch headers. These being just the major visible valves, all operated by manual hand wheels, no automation. The four inch headers feed each process tube through 2,004 individual pigtails. Each inlet having an orifice to provide a flow index. Original water flow was 35,000 GPM with 80,000 GPM in its later upgrades. Since water also reduces the reaction by absorbing neutrons, the water jacket thickness was only 0.086 inch. The entire reactor water volume within the 28 ft process tubes was only 400 gallons giving a dwell time of water in the reactor ranging ~0.8 to ~0.3 seconds for average flow velocities of 35-90 ft/sec. The water temperature would rise from river ambient temperature to 195ºF. Compared to a modern power reactor, this permitted non-pressurized single pass cooling conceding the simplest development to achieve

Tripping Over Engineering: Going Nuclear

the quickest production.

While there is distributed detection on neutron flux and temperatures, these tend to provide for production improvement and diagnostics. With a process tube outlet pressure at ambient, the peak water temperature is approaching boiling. Upon flashing to steam the fuel charges would have significantly less cooling.

When built, it was unknown whether the fissile rate would increase or decrease due to the loss of the water as a neutron absorber. Whether the loss of cooling would lead to a temperature excursion that would: ignite the aluminum fuel jackets, melt the jackets and ignite the uranium fuel or ignite the graphite moderator blocks, leaving the area a Chernobyl. Therefore, the most critical instrumentation was that which showed the cooling system working as advertised. Each one of the 2,004 process ports had a hard plumbed instrument tubing connection to the control room.

This tubing connected to 2,004 individual pressure gages with high and low trip settings. A rise in pressure could indicate fouling or obstruction of the water gap. A drop in pressure indicating a reduction in inlet flow or a rupture in a process tube. Deviation of any process tube pressure from allowable would directly SCRAM the reactor. The clockwork simplicity is posted above the panel "CAUTION: BUMPING PANEL MAY CAUSE SCRAM". Supporting this tubing are 2,004 three way bleed and purge valves feeding back to the instrument tubing to control any fouling. In the war time configuration, the bleed purge was served by 4,008 needle valves. Some operator had to squeeze between the racks, bleed and purge, bleed and purge, bleed and purge......

The valve pit displays an expanse of catwalk over line of gate valves mating with checks and strainers. These route an array of twelve inch to twenty inch inlet lines to the two thirty-six inch headers feeding the reactor. Supplies ranged from the multiple electric and steam pumps to an elevated gravity tank. To provide for the most immediate response, the valves are run open with flow management by check valve. Only three electric operators are seen for all of these valves.

Tripping Over Engineering: Going Nuclear

The valves are conspicuously common. They were no different from those going into any other plant at the time or now for that matter, handling ambient temperature water below 350 PSI. In many cases even construction workers didn't know the specifics of what they were creating. Other than being given a materials allocation and a specific delivery priority requirement, the suppliers had no idea of the contribution that they made to the war effort. These valves are not the N-Stamped exhaustingly documented valves we see in modern reactors. As an old boss once said, you can't ship an N-Stamp valve until the weight of the paperwork equals the weight of the valve.

E.I. Dupont de Nemours had a level of experience in constructing large war time chemical plants and was approached for construction of the reactors. Considering that they had no experience in nuclear operations, nor did anyone else in the world, they were hesitant to accept the project. Since they were one of the few capable of such a project, the government waived all responsibility of the contractor. In many cases the reactor was constructed with the nuclear calculations and conceptual drawings. Once the basic set of criteria were established, there was an elegant simplicity.

On 26 September 1944, dry criticality was tested at a very low power and on 27 September after midnight it went critical with full cooling. Early that evening, after continuously withdrawing control rods, reactivity stopped. Early the next morning it spontaneously restarted. Twelve hours later, fission again ground to a halt. While there was fear that boron might be scaling from the water which could severely alter planned operation, it was determined by the power decline that Xenon byproduct isotopes were trapping the neutrons and decay permitting restored fission. Dupont had the chemical plant experience of understanding that processes are never ideal and it is always useful to have a cushion. The initial reactor design had specified a 1,500 process tube cylindrical reactor core. Dupont had held out for squaring out the core with an additional 504 process tubes and the additional drilling time. This additional reactivity overpowered the Xenon poisoning and permitted the reactor to operate at designed power. You should always listen to

Tripping Over Engineering: Going Nuclear

the process engineers {Rhodes 1986}.

Regrettably, most auxiliary facilities have been demolished. Pump houses were powered by direct power lines from the Bonneville and Grand Cooley Dams with a steam plant and its six month's of coal in reserve. Water treatment filtered, deaerated and acid treated to protect the reactor. Effluent water management neutralized acidity and decayed short life isotopes.

Many very important systems are not readily visible. The original "last ditch" SCRAM system was three pressurized tanks holding a borate liquid solution. A set of quick acting valves would flood the safety rod tubes with this solution to kill the fissioning. While this had significant reliability, there were risks in trapping borate in areas that couldn't be cleared, killing portions of the reactor. This was replaced by a bin with hundreds of boron balls dumping into channels. After SCRAM, valve under the reactor could be opened and the balls simply dumped out, literally a "ball" valve. A helium circulation system provided dry cooling to the graphite stack. While the entirely new presence of nuclear science and nuclear engineering get their due, extensively typical industrial equipment kept it working.

Six reactors were initially estimated to provide the fissile capacity to generate plutonium at the rate necessary for weapons production. By the time construction started, improved capacity calculations and the change to an implosion device from the original gun design permitted the reduction to three reactors B, D & F with much greater safety separation along the Columbia River. Space had been allocated so the survivors could endure regardless of an individual reactor catastrophe and abandonment.

B reactor was built with a three year intended production life. One year from ground breaking to reactor criticality. One and a half years of war time production. Two years idle then resuming operation for another twenty years. It was the first industrial scale nuclear reactor and educated all that followed. The three sisters initially provided plutonium for the Gadget (1st atomic detonation) and Fat Man (3rd atomic detonation, 2nd and last atomic detonation in warfare). Eventually providing 57 tons of plutonium for most of the US plutonium fueled tests and weapons.

Two other historic points of Hanford facility are notable. The systems for EXTRACTION of plutonium actually cost more than the reactors. The Queen Marys 221T/U/B were 800 ft long, 65 ft wide, 80 ft high with forty process pools each. Also called the canyons, extensive and nasty chemical processes handling radioactive materials meant that once running, all operation and maintenance had to be done with remote control through periscopes and television. This is a simplified process, the details get painful. In the 221T/U/B: sodium hydroxide strips aluminum coating, nitric acid and mercury catalyst prepares it for multi stage bismuth phosphate precipitation. In the 224T/B: precipitate is treated for lanthanum fluoride precipitation which is treated with potassium hydroxide. In the 221Z: peroxide precipitation resulting in plutonium nitrate. Shipped to Los Alamos for final plutonium metal reduction. There have been, on occasion, tours of the T Plant, if you are lucky enough to see it, congratulations.

For those interested, there are significant declassified construction documents showing the progress and planning for the facilities. Typical of military expedient facilities, they were built with production as the only requirement and extensive residue to be addressed at some later time. A number of chemical weapons and explosives plants had similar issues, except for the problem of radioactivity. Extensive expense is now redressing the decisions of the time. There have been occasional public tours of some of the clean up sites.
{DOE 1943}

Tripping Over Engineering: Going Nuclear

Fireset

Manhattan Project
Los Alamos NM [BSM]
Service 1945-1950
Ignition system used on all Fat Man weapons

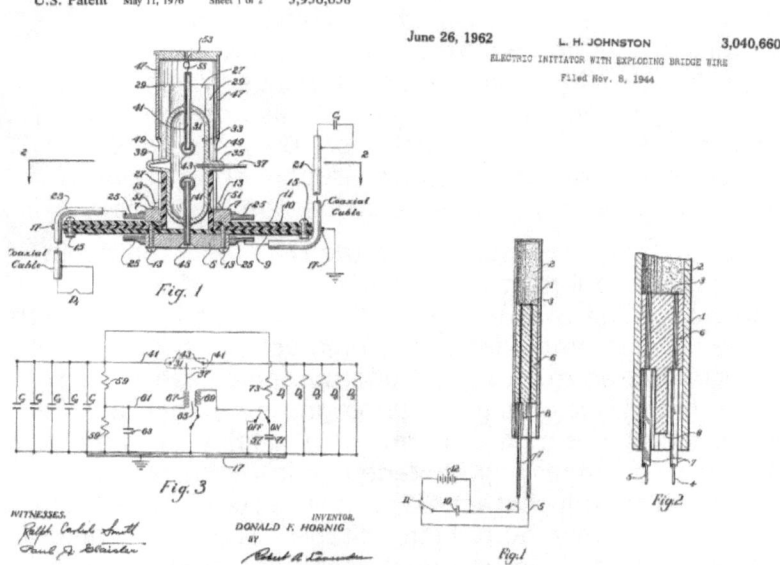

Patents for Spark Gap Trigger and Exploding Bridgewire {USPTO}.

Little Boy was a uranium gun type weapon. Gun type weapons have a particular design advantage, they are a gun. A barrel with propellant, primer and hundreds of years of development. No timing or precision issues, fire it and off it goes. The reactor production of plutonium created a significant concentration of Pu240 which would predetonate in a gun device mandating the implosion device of the Gadget and Fat Man.

Unlike the gun where the two masses are simply smashed together from a single point charge, an implosion device demands more dispersed masses. This was achieved with plutonium hemispheres which are concentrically compressed onto a polonium-beryllium initiator. The propagation of the wave front in an explosive blanket lacks adequate sphericity.

Tripping Over Engineering: Going Nuclear

The best distribution of compression was found to be with a 32 evenly distributed ignitors around a multilayer explosive charge. In the end, you are trying to concentrically crush the fissile material toward a convergent point. Any asymmetry in distribution or trigger timing would result in a dud.

This was problematic with a typical blasting cap or primer having variable heating to variable ignition with a variable burn rate. The solution was the "exploding bridgewire". With a high enough voltage and current, hit all of the bridgewires providing almost simultaneous priming. The high power requirement of the bridgewires also provided some safety from stray current situations. But, this system required two more items, a power supply of appropriate characteristics and a method to switch the high voltage.

Sitting in [BSM] is an item called the "Fireset". Found in storage in Albuquerque with a serial number of early production. The consumption rate of Firesets infers that this may be the last of its kind. The high voltage power supply was reconfigured from a giant photographic flash system designed for night time aerial photography. That left the need to switch the high voltage. The problem with a typical contact switch under high voltage is a lower current arc bridging before full contact. The real beauty was in Donald Hornig's understanding of high voltage. A gap between electrodes can be held to a relatively high voltage if an ionization path does not initiate. The use of an additional electrode which while at a smaller gap, uses a lower energy trigger to initiate a discharge path to one of the electrodes. The ionization path shortens the spark gap for the main electrodes and boom, major discharge.

Although this ignition system had specific tolerances for stray voltage, there were conditions that were potentially catastrophic. There were concerns over thunderstorms in the final setting of the Gadget at Trinity. A week before, a test X-unit at the site was triggered by static during a lightning storm {Schlosser 2013}.

Spark gap triggers are still in use today for rapid switching of high voltages. You will see a lot of weapons casings, but this is one of the few actual critical nuclear working parts that you will find. And you have to admit, for what it does, even the name Fireset is sort of cool.

Tripping Over Engineering: Going Nuclear

Trinity

Manhattan Project
White Sands Missile Range NM [WSMR]
16 July 1945
1st atomic detonation

Trinity - Name selected by Robert Oppenheimer, inspired by the poems of John Donne which addressed the paradox of death and redemption.

The Gadget in the tower {DOE}.

Little Boy was a gun type U235 weapon. They were so confident in its performance that they felt no need to even test before deployment. The physics and control of an implosion weapon was much more tedious and complex. There was a significant concern of a dud, requiring an actual test detonation. The Gadget was the first atomic detonation, the first implosion type device and the first plutonium device.

Tripping Over Engineering: Going Nuclear

An old colonial Spanish trail that was used to survey the area for the intended test was Jornada del Muerto, literally, journey of the dead {Rhodes 1986}.

At the actual ground zero, it's somewhat counter intuitive when you arrive to see just a fence around the space and the obelisk in the center. Although it's been cleaned up significantly since the test you would still expect to see much more notable damage around. There is a shed that used to have a window through which you could see the original trinitite. This window is no longer opened because the trinitite is sanded over and no longer visible. There is the remnant of an anchor for the original shot tower.

Given that this would be the first nuclear detonation in the world, there was particular concern for the data which could be acquired during the test. A wood tower was built to support a hemispherical pile of 100 tons of TNT to best simulate the blast propagation. A system of tubing distributed a radioactive tracer over the stack. It was fired on 7 May with extensive instrumentation. This gave a proper 0.10 KtE index yield, and maybe the concept of TNT equivalent. The Gadget detonated at 20 KtE.

There is a gadget replica at [NMNSH]. One of the remaining Tinian personnel worked on the replica. The seven piece aluminum sphere is an actual implosion shell from the MK-3 sitting next to their B-29. While I assumed that it is the postwar production, my fuzzy photos look like a foundry date of 4-23-45 making it a wartime shell.

The MacDonald ranch house and the remaining Trinity structures are amazing in that they were only 2 miles from detonation and received limited damage. The final assembly of the gadget was performed at the ranch house with plastic hung around the walls to reduce dust. There remain notes written on the walls and the original trim painting inside the house.

A relic of the test is the carcass of Jumbo. Ten foot diameter, twenty five feet long and 194 tons with a 50,000 psi design pressure. At the time, the heaviest rail shipped single item. It was a Babcock & Wilcox built pressure vessel intended to

Tripping Over Engineering: Going Nuclear

contain the gadget explosion, at least the chemical explosion. It was originally intended to fire the Gadget inside Jumbo. In the event of a dud, the plutonium would be contained along with the chemical blast. Extracting thirteen and a half pounds of plutonium blasted into the steel shell would have been an interesting prospect unto itself. Another concern would have been the event of a fizzle with 5,300 lbs of chemical explosive and a 10-20 tE fission predetonation. Some fissioning, plenty of radioactivity and incomplete containment. But, confidence in the gadget performance and concern over loss of data acquisition from Jumbo interference lead to its desertion. Eight 500 pound bombs were eventually attempted to destroy Jumbo, it did blow off the ends with the bombs laying on the side. I would assume that the gadget with a mix of explosive velocities and centered in Jumbo may have been properly contained. [BSM] has an interesting piece from clean up activities a Los Alamos. They found several Jumbinos, explosively tested scale models of Jumbo.

Less than 27 miles from ground zero is San Antonio NM, where we stayed at an RV park along the Rio Grande. We were warmly greeted by the owner who had seen the flash and remembered the boom from when she was twelve. The Army put out a statement that it was the explosion of a munitions train, but many people had believed that it was something far more significant. The viewing of the Trinity site as a White Sands open house is available twice a year. An open house the Very Large Array is run on the same dates. Both areas are must see.

Tripping Over Engineering: Going Nuclear

Fermiac
**Manhattan Project Postwar
Los Alamos NM [BSM]
Used 1946 - 1948
Mechanical graphic replacement for digital computer**

Stanislaw Ulam holding Fermiac {LANL}.

When Erico Fermi reached criticality with CP-1, he pulled out his slide rule for some calculations and announced the self sustaining reaction. Calculations for the first weapons were performed in a predigital world with a lot of number crunching to turn theoretical into a producible device. The noun for computer at the time related to a person who did the math

Tripping Over Engineering: Going Nuclear

and not the device. The calculations at Los Alamos were performed predominately by staffs of young women, with the pressure to limit additional personnel for security reasons, including the wives of Los Alamos staff. There were mechanical differential analyzers and punch card calculators, but often calculations were iteratively performed by banging away on Marchant mechanical calculators one keypunch at a time.

A wartime designed digital device did reach operation after the war. Eniac had been developed to run artillery ballistics tables, a brute force repetitive calculation. Surreptitiously, its first calculation was for the Teller Super, the hydrogen bomb. Monte Carlo Simulation has much of its provenance with Los Alamos mathematicians and physicists. While Los Alamos brought about great computing developments, there were times that expedients had to be found.

Enrico Fermi wanted a way of simulating neutron paths through materials while Eniac was being moved. Only one device is known to exist, in service for about two years, made by a colleague of Fermi. It is commonly referred to as Fermiac. A Monte Carlo method neutron transport graphic analog computer. It is a small brass frame with a set of wheels and an acrylic plate. It would roll across a scale drawing of the device leaving a pencil trail for the neutron with ratios set for the material and a random selection for fast or slow neutrons. Long before digital systems provided good graphic outputs, Fermiac provided an enchanting graphic view of statistical neutron movement. [BSM] also has a spectrum of the hottest computing equipment from Eniac to Roadrunner.

Tripping Over Engineering: Going Nuclear

Redstone Arsenal
Huntsville AL   [USSRC]
1949 to present
Birthplace of the US space program

Jupiter C on Redstone Test Stand {LOC}.

Tripping Over Engineering: Going Nuclear

When Operation Paper Clip recruited hundreds of German scientists and technicians at the end of the war, the intended headquarters for rocket research was out in the desert of Fort Bliss due to proximity of White Sands Missile Range. There was a push back by the Germans to move to a location which would have more recognizable weather. A former Army chemical weapons plant and depot near the Tennessee River in northern Alabama provided adequate facilities and environment for rocket development. Huntsville Arsenal eventually became Redstone Arsenal from the red clay soils typical in the region. And in the post war base closings, Redstone received another lease on life with the Ordnance Rocket Center. Parts of the facility continue service as Marshall Space Flight Center and other NASA functions. It also continues as the center of Army missile programs and a range of Army logistics, aviation and space programs.

While there had been wartime progress on a range of rockets, the appropriated V-2s provided an instant field tested weapon with a 2,000 lb warhead and a range of almost 200 miles. It was even guided in variants with passive radio navigation, command guidance and rudimentary inertial guidance. The V-2 developed a range control system that would be the ballistic missile standard for decades. After heaving the missile to its free fall range, simply shut off the engine and coast to target. At the time, the ultimate non-interceptable weapon system.

During the Korean War, a program was started to launch one of the large thermonuclear warheads out to 500 miles. The Army had been learning from the V-2 missiles at White Sands with some improvements, but a distinct leap was needed. To test this major step at Redstone and stay under budget visibility, the Interim Test Stand was built for $25,000 with an extensive array of creatively salvaged materials. In order to have a protected area for operations and instrumentation, three railroad tank cars were buried nearby. Periscopes provided observation and filming in the event a blast scalped the bunkers. Given the "interim" moniker in 1952, it was upgraded into the 1960s for testing newer missiles.

While the Redstone missile never achieved the initial range

Tripping Over Engineering: Going Nuclear

targets, it did achieve a significant warhead lift capability. Its V-2 heritage was obvious in the control method of carbon jet vanes and aero rudders. The motor was distinctly different with the plate injector, but it retained the alcohol/water and LOx fuel system. While the 25% water in the alcohol distinctly cut into the energy capability, it did provide for cooling of the motor and nozzle. Extensive experience in handling LOx and eventually hydrazine laid the ground work for modern rocketry. Besides the Redstone missile, derivatives tested at this location launched the first US astronaut into space and the first US orbital satellite.

Engine and test stands were added for Jupiter and Saturn systems. While the Redstone missile itself was not particularly useful for its original design intent, it provided a lift capability to boost the US into the Space Race. See PGM-11 Redstone.

Tripping Over Engineering: Going Nuclear

Nuclear Testing
Nevada National Security Site [NNSS]
National Atomic Testing Museum [NATM]
Nuclear testing 1951-1992
928 nuclear detonation tests

Operation Buster-Jangle Easy shot 5 November 1951.   Mk-7 freefall prototype from B-50 {DTRA}.

Ninety miles North West of Las Vegas is a location once simply known as the Nevada Test Site. From 1951-1992, 928 total official nuclear test detonations were performed. 100 above ground, 828 sub surface. Some of these were simultaneous detonations within one test, so there were actually more devices fired. The US performed an additional 104 tests, primarily in the South Pacific. The Soviets performed 727 tests including the 50 MtE Tsar Bomba, largest nuclear detonation ever. The Soviets performed 456 of their tests at the Sempalatinsk Test Site, aka The Polygon in Kazakhstan. Add other European and Asian tests, a world total of 2,121 tests with 2,476 devices, including 188 nuclear detonations of

Tripping Over Engineering: Going Nuclear

"Peaceful Purposes".

Some tests were very significant from a technical standpoint, some cases appear to be timed and publicized for maximum propaganda. Just the extent of testing indicated to each side how much yield and unequivocal production the other could achieve. In the South Pacific tests, the presence of Soviet trawlers was well known. Underground testing is readily detected on seismic. There was even a secret project to detonate a W25 warhead from the AIR-2 Genie on the moon. The Air Force ICBM program was expected to have an adequate launch capability by 1959 which tends to lean toward an Atlas based launch vehicle. There was some science to be accomplished along with a whole lot of, "how do you like them apples" in the Soviets face. This project almost went away quietly except for a biographer finding the titles of two papers in notes from a young doctoral student working on the project, Carl Sagan.

But, this is a digression from the point at hand. In order to perform each of these tests, in particular sub surface, an extensive amount of engineering and construction effort is required. Lots of geology surveys, lots of drilling and digging, lots of instrumentation, miles and miles of wire for each test. The [NATM] extensively shows the effort that goes into this testing, including a range of instrumentation from actual tests.

The Nevada National Security Site [NNSS] has an annual tour season which books up quickly.
Things you will see no where else:
Frenchman Flat, used for extensive above ground tests (my favorite, Upshot Knothole Grable with M65 Atomic Annie).
Sedan Crater (big hole in ground from "peaceful use" test). This was from the "Operation Plowshare" program to demonstrate the great usefulness and practicality of using nuclear warheads for civilian massive earthmoving projects. There was actually a consideration that the construction of canals and lakes could be more effectively performed with nuclear devices.
Apple II houses (from 2nd televised test and public effects demonstration). These demonstrations are well known from extensive use of the video in movies. There is a

Tripping Over Engineering: Going Nuclear

question of whether this showed survivability or fatalism. The intent was to show destructiveness and encourage personal defense efforts for survival planning. Along with John Hersey's article and book on Hiroshima, I would suspect that in a lot of cases, these demonstrations provoked more a sense of futility than survivability.
Nuclear waste management installations.
Facilities for evaluation and training for nuclear events.
Besides the listed areas, I understand that there are also extensive test relics along the tour routes.

There were other test detonations away from NNSS that raise more questions than they answer. The current method of hydraulic fracturing, pushing a water sand slurry into a pay zone for improved production has been used since the 1950s, there was another option tested in 1967 on Project Gasbuggy. In an effort to enhance gas production, the AEC funded the test shot of a 29 KtE nuclear device 4,227 ft underground in the Carson National Forest of New Mexico. Another gas fracturing in 1969 was the Project Rulison with a 40 KtE device down 8,400 ft near Parachute Colorado. While enhancing gas flow, strangely enough, the gas had significant radioactive contamination, so why not do one more test in 1973. Project Rio Blanco, roughly centered between Rangely and Rifle Colorado, was performed using three 33KtE devices fired simultaneously at 5,800 ft, 6,200 ft and 6,700 ft deep. While all of these tests were cleaned up at the surface and have specific drilling restrictions in the area, they are all marked and are publicly accessible {DOE 2006} {DOE 2015}.

EBR-1
Idaho National Lab [INL]
Operated 1951 - 1964
20 Dec 1951 1st nuclear powered usable electricity
EBR-1 - Experimental Breeder Reactor One

EBR-1 start, reactor building to right {ANL}.

EBR-1 was originally named Chicago Pile 4 as the location was initially referred to as Argonne West (compared to Chicago, way west). A more interesting internal name was ZIP for physicist Walter Zinn's Infernal Pile. A fast breeder reactor provides much higher conversion efficiency of fissile material compared to a light water reactor. It can also breed fissile isotopes with fast neutrons rather than moderating them to provide neutron capture. EBR-1 was the first NaK cooled and second fast breeder reactor, only Clementine preceding it as the first liquid metal reactor and fast breeder.

Rather than use water as primary coolant which required pressurization of the reactor, a liquid metal of molten sodium and potassium (NaK) was pooled around the core. This

Tripping Over Engineering: Going Nuclear

reduced pressure risk of the reactor and eliminated steam flashing in the event of a loss of reactor feed containment since NaK is liquid up to 1445°F. Full power shut downs demonstrated a significant natural safety advantage under high energy conditions.

But, there is also the dark side to NaK. If it leaks into an air environment, it spontaneously burns at extremely high rates. If water is applied to it, it accelerates the combustion. Beyond the primary coolant loop of the reactor, it had a quite conventional steam system. So there is always the risk of high pressure steam in the boiler with low pressure NaK on the other side. Any leak is a concern as experienced in 1995 on the Japanese Monju reactor.

While NaK circulation was initially done with conventional pumps, it was always problematic for shaft seals. But being a liquid metal, the magnetic characteristics let you do some weird things. By providing a set of coils around a magnetically porous pipe and feeding it an alternating current, the magnetic field simply motors the NaK along.

On 20 December 1951 EBR-1 lit up four 200 watt light bulbs to be the first reactor designed to generate electricity. That's right, four bulbs, and remember that these aren't modern low energy bulbs. So proof of production was a massive 0.8 KW. The following day it produced 100 KW. It typically had only a 200 KW electric capability, providing its own power until decommissioned in 1964. Considering that previous reactors were prodigious CONSUMERS of power, even self supporting power was a notable development. Later it would generate the first electricity fueled by plutonium.

Public nuclear power would not occur until 17 July 1955 when nearby Arco was powered by the BORAX-III reactor. Its predecessor BORAX-I was intentionally taken to prompt critical, blowing out the top of the reactor to test damage and safety under deviations from design operation. In some ways INL was used like the NTS except that it intentionally damaged reactors instead of detonating weapons. Over fifty experimental reactors have been operated at INL.

Liquid metal fast reactors were of adequate interest that the second nuclear submarine, USS Seawolf (SSN-575) was built with a similar reactor. It was considered far more advanced than the light water reactor in the Nautilus, bringing an array of advanced problems. After only a couple of operational years, it spent two years having the reactor replaced by a more conventional light water reactor. Rickover hated the liquid metal reactor and stated that if the world were made of oceans of NaK, somebody would want to use water.

Fast breeder reactors became less economical with more available uranium. There was also concern on production of weapons grade materials. Particularly since plutonium remained a quicker approach than uranium enrichment at the time. With modern enrichment, and a rogue state just wanting "A" nuclear device to join the club, the weapons material argument against fast breeders is probably far less valid than it once was. This led to its decline in the US, with other countries continuing to develop fast breeders.

DELIVERY

In the beginning of nuclear weapons, weapon delivery was very simple. A warhead WAS a free fall bomb and the only reasonable delivery system WAS a heavy bomber. In some cases a particular bomb was restricted not only to a single type of bomber, but a single configuration of that model.

Eventually, practically every type of weapon delivery system achieved a nuclear warhead capability. Fortunately, out of all of those warheads, no one ever initiated a strike beyond the first two bombs or decided to initiate armageddon on their own. There were close calls over the years, in some cases with one individual being the only force of reason.

For the 1980s, the Soviets and US agreed to limit strategic weapons capability to 2,400 delivery vehicles per side. For the US this translated to 440 bombers, 1,054 land based ICBMs and 896 submarine launched ballistic missiles {Polmar 1975}. But this did not address "tactical" nuclear weapons which were higher yield than Fat Man and could be carried by practically any aircraft. It also left flexibility in multiple reentry vehicles on the latest missiles and the planned 240 unit B-1 fleet with 24 bombs or cruise missiles on each aircraft. In some ways you have to wonder if the plan for arms limitations is purely an effort to limit expenses by both sides, because it looks like there was still a significant devastation capability.

Tripping Over Engineering: Going Nuclear

BOMBS

When bombs were the ONLY nuclear weapons, development proceeded two directions, higher yield and more portable packaging. Higher yield was expected from the wartime Teller concepts, in the interim accepting that these devices would be huge.

Over time, higher yield became smaller while aircraft developed more load capacity and industrialization provided large numbers of production line nukes. The original airburst area bombs developed a range of target capabilities. Bombs could now operate as airburst, contact, delayed laydown or hydrostatic, creating all purpose land and sea killing machines.

Depending on how you want to argue it, there may be roughly 35 bomb models that have been deployed in the US. Each with several levels of upgrades, several methods of application, several warhead derivatives from the original device and in some cases selective nuclear capability. Some of the early significant bombs are shown.

Tripping Over Engineering: Going Nuclear

Mk-1 Little Boy
Manhattan Project
Hiroshima Japan
6 August 1945
1st nuclear use in warfare

Little Boy - an ongoing created name from the Fat Man/Thin Man naming of the book series.

Little Boy in loading pit on Tinian, note bomb bay door top right {NAC}.

Tripping Over Engineering: Going Nuclear

A gun-type weapon permits a simplicity of design requiring brute force of operation. The key is to carry two subcritical masses adequately separated in distance to prevent criticality and cramming them together fast and long enough to go supercritical. Too low of velocity and it will go prompt critical, blowing apart the two masses before supercriticality. The quick solution was to have one mass fixed with another mass up a barrel that can achieve the required velocities.

The initial Thin Man Mk-2 plutonium gun device required a bomb length of 17 ft and 8,000 lbs with 75 ksi chamber pressure and 3,000 ft/sec muzzle velocity. Pure Pu239 would have allowed this design, but with more intensive production there was also more content of another plutonium, Pu240. Criticality tests found Pu240 spontaneous fission to be much more active than Pu239, preventing any reasonable gun device.

With U235, the level of spontaneous fission permitted a much smaller gun of 6 ft long and 1,000 lbs providing 1,000 ft/sec in a 10 ft long 9,700 lb bomb {Polmar 2009}. This was much more compact and aerodynamically better behaved bomb than Thin Man. But gun devices naturally have limits to efficiency. Little boy contained practically all of the enriched U235 then in existence, 141 lbs. At detonation, only ~1.38% fissioned, physically blowing the remaining 98.62% apart generating a yield of 15 KtE {Schlosser 2013}.

The simplicity of firing also brought a simplicity of accidental detonation, since a gun only needs the primer fired. So it was decided by the armorer not to install the four cordite firing charges until airborne since, any fire or possibly static or impact could detonate. These charges were fired by any one of three standard Navy primers, propelling six U235 rings down the barrel to encompass six lower U235 target rings with polonium-beryllium initiators. Holding the target under these loads was a tungsten-carbide anvil. Once the charges were loaded, all that remained were removing three safe plugs and installing three armed plugs. One of the green safety plugs pulled before the Hiroshima bombing is at [NMUSN]. A total of eight units were built for testing. These were used for handling and fuse tests, several being dropped around Tinian to confirm function.

Tripping Over Engineering: Going Nuclear

The fusing system had a timer that locked out the radar altimeter for fifteen seconds to prevent triggering from aircraft detection. A barometric switch would also lock out the radar until dropping below 6,600 ft. The radar altimeter was modified from existing Archie tail warning proximity radar firing at 1,900 ft above ground level.

Little Boy was the first functional gun-type weapon and the second nuclear detonation in the world. Lacking the efficiency of an implosion device, it was not intended for use beyond the Hiroshima bomb. Postwar, Hanford had production limits due to the Wigner effect so six more were built. There were limited assembly notes or even drawings. They located original Little Boy diaspora to recreate the two year old design. A discussion with a machinist on a particular aluminum tube found that he had wrapped sheet around Coke bottles to achieve the shape {Schlosser 2013}. The size limits of the P-2V Neptune brought about the production of an additional 25 since it could not fit a Fat Man in the bomb bay. But, apparently there was only nuclear material for six bombs. These were fully withdrawn from service in January 1951. {Sublette}

While there are replicas in many locations, actual inerted production bombs are at [NMUSAF] & [NMUSN]. Many "replicas" on display may be the Navy 25, particularly since the [NMNSH] has anchor marks next to panel part numbers.

Tripping Over Engineering: Going Nuclear

Mk-3 Fat Man
Manhattan Project
Nagasaki Japan
9 August 1945
2nd and last nuclear use in warfare
Fat Man - A character in The Thin Man book series

Crossroads Baker submerged test 25 July 1946 {NHHC}.

As an insurance policy, there were multiple bomb designs running simultaneously. With the discovery of high Pu240 in heavily reacted plutonium, the implosion device was no longer an alternate, it was mandatory. Multiple compression methods were reviewed before the explosive lens compression of a plutonium shell was achieved.

In order to get the ignition simultaneous, the Fireset and exploding bridgewire was developed. Now all that was necessary was getting multiple convex shock waves to bend into a spherical shock wave while propagating through the explosive. Experience with armor piercing shaped charges demonstrated the shaping of explosive forces. Directly under the detonator between layers of high velocity explosive, there was a convex section of lower velocity explosive so that the initial convex shock wave would turn inside out before it hit the next section of high velocity explosive. By the time all of the explosive was fired, the plutonium core received spherical

Tripping Over Engineering: Going Nuclear

compression. The cost of this beautiful concept, 11 ft long, 5 ft diameter and 10,300 lbs with half of the weight in the 17 inch thick 5,300 lb mosaic of high explosives {Polmar 2009}.

There was a tedious complexity to the device, the explosive lenses had an amazingly high scrap rate, eventually using a dental drill to access cavities and hand pour molten explosive to repair {Rhodes 1986}. Even Oppenheimer called the Mark 3 a "haywire contraption" {Schlosser 2013}. With the rush to get weapons to Tinian, there were a range of problems from incorrect wiring requiring cutting and soldering during final assembly to false armed lights requiring debugging in flight {Schlosser 2013}. But, Fat Man showed one of the key benefits to the implosion device, compared to the inefficiency of Little Boy, ~20% fission from 13.5 lbs of plutonium generating 21 KtE. Despite the problems, it was the third nuclear detonation in the world.

In the post war years the Mk-3 was considered such a high risk that they would not fly an assembled bomb in the US, but that did not stop them from flying over other countries, even allies. At the start of the Cold War, the Mk-3 was the nuclear capability, in 1948 the US had enough parts to build 56 complete bombs {Schlosser 2013}. As an operational ready weapon, the bomb had to be disassembled every nine days to replace the batteries which required two rechargings during that period. The plutonium core had to be removed after ten days to prevent alpha particle damage to components {Polmar 2009}.

Besides Nagasaki, Fat Man devices were tested on Crossroads Able and Baker shots. These are probably the most publicly documented and recognized of the nuclear tests. There were eventually 120 Mk-3s built, in service until 1950. While the Fat Man family of devices was developed for the Army Air Force, Navy use of the AJ Savage gave the space and weight capacities for their strategic use from carriers. The Mk-3 derivatives were major advances from wartime expedient weapons to safer and less maintenance intensive devices. This gave the derivatives of the Mk-3 far longer service lives.

There are a lot of Fat Man "replicas", most of them casings from the 120 built post war. These are all Mk-3 Fat Man devices with improvements as could be integrated in the original casing. These were actual weapons, they just weren't the one that was dropped. There are casings at [AFAM], [NMUSAF] & [BSM]. [WSMR] has one it displays at the Trinity open house, [NMNSH] has one sitting next to a B-29, one inside and a Mk-3 implosion package shown as a Gadget replica, see Trinity.

Fat Man Derivatives

Mk-4, 550 bombs, service 1949-1953: While it was generally the same size and weight of the Mk-3, it had greatly improved logistic and safety advantages. The Mark 4 permitted in flight insertion of the core in flight through a nose hatch and did not charge the X-unit until after drop. This method of insertion mandated adequate access to the nose of the bomb in flight. The composite uranium/plutonium core allowed easier assembly and at least a couple of weeks reliable standby {Schlosser 2013}. There were a range of design yields from 1 KtE up to 31KtE.

Mk-5, 140 bombs, service 1952-1963: While the Mk-5 was considered to be a new design, it was the first with auto in flight insertion on an implosion bomb. This allowed the core to be carried in a safe position in the casing. An external terminated cable provided for setting the core without requiring open access to the casing. This allowed use of more compact bombers without flight accessible bomb bays. It had greatly improved size and yield with 44 inch diameter, 3,175 lbs and up to 120 KtE. Ignition was increased to a 92 point system. It was used as the fission trigger for the first thermonuclear device test. [NMNSH] has one.

Mk-6, 1,100 bombs, service 1951-1962: This was a continued improvement of the Mk-4. The Mk-6 initially continued the 32 point ignition system, later using a 60 point system. While overall dimensions were the same as previous, it did achieve a weight reduction to 8,500 lbs and yield up to 120 KtE. [NMUSAF] has one. [MOA] has one sitting under a B-29 in loading position.

Mk-13/18, 90 bombs, service 1953-1956: While these both were Mk-6 derivatives, they were starting to get significantly different. The Mk-13 used a 92 point ignition. Then the Mk-18 used a large highly enriched uranium core. Mk-18 was tested on Ivy King at the highest pure fission detonation, 500 KtE. The development of thermonuclear weapons had greatly outpaced the yield of fission devices. Although ninety units were in service, the four mass configuration was particularly hazardous. The firing of any single detonator risked a prompt critical event. In deployment, an aluminum/boron chain was carried in the pit which was removed while arming. Due to this risk, they were modified down to Mk-6 configurations in their later service.

Mk-7

Service 1952-1968 Air Force & Navy
1700+ produced
1st tactical nuclear bomb
Every pilot's nuke

The Mk-7 was considered the first general purpose fleetwide tactical nuclear bomb. Thirty inch diameter, fifteen ft long and 1,700 lbs with streamlining for external carriage. The lower fin also folded to give a bit more clearance on take off and landing. Most fighters and strike aircraft had adequate load capability to carry the Mk-7, but hardpoints and ground clearance continued to restrict some aircraft. A comparative 2,000 lb general purpose bomb, the M66 was 10 ft long, 23 inch diameter with 32 inch across the tail. While there were specific aircraft rated to carry the Mk-7 with appropriate bombing systems, it is easy to see that the concurrence of weapon and aircraft technology meant that if there was a full scale nuclear exchange, there was an extensive fleet that COULD carry it. Gone were the days of bomber only nuclear capability.

The much lighter warhead also provided the option of nuclear missiles using the W7. This warhead was used in Honest John and Corporal missiles, Betty depth charge and BOAR rocket. It was one of the earliest weapons with variable yield, from 8 to 61 KtE, eventually with PAL security. It was still a multipoint implosion type fission warhead, but with a 92 point detonator system and U235.

The Mk-7 had two nuclear tests, both B-50 air drops. Buster Jangle Easy on 5 November 1951 tested a Mk-7 prototype. Tumbler Snapper Dog on 1 May 1952 tested a Mk-7 using an early tritium boosting design.

The Mk-7 also became the first US air to surface nuclear rocket also making it the first nuclear bomber standoff weapon. The BOAR rocket deserves a bit of explanation due to a range of issues. 225 were in service from 1956 to 1963, not bad considering that it was intended to be a three year interim weapon. It was 2,000 lbs and similar external dimensions to the Mk-7. With only a 7.5 mile range,

Tripping Over Engineering: Going Nuclear

essentially a Mk-7 with a rocket kicker. The need for this fairly low performance was the change in tactics due to the Mk-7 itself. Where prior weapons were delivered from high altitudes at the highest speeds achievable, delivering big bombs with parachutes gave strategic bombers a much larger blast avoidance and escape window. Fighter and strike aircraft were best determined to attack from low altitudes. In order to get some separation at the time of detonation, the Low Altitude Bombing System (LABS) was used for delivery. The BOAR was operated in the same manner, with the rocket lobbing at least a hopeful separation between the aircraft and the detonation. Apparently, there was even an air-to-air version of the BOAR considered for engaging bomber formations, a cheap AIR-2 Genie {Silverstone 2013}. There is a BOAR at [USNMAT]. There are Mk-7s at [AMSE], [NMNSH] & [NMUSAF]. {Michel 2003}

As Big As You Can Carry

Mk-16/TX-16/EC-16
5 bombs built January 1954
Removed from service by mid 1954
1st deployed thermonuclear weapon
Only cryogenic nuclear weapon

Ivy Mike was the first test of a true thermonuclear device on 1 November 1952. Due to the use of cryogenic deuterium for the fusion reaction, the "installation" weighed 82 tons with a liquid hydrogen plant for cooling. Since a fission/fusion device provided the step up to MtE yield levels, it was decided to proceed with a bomb design. This brought life to Edward Teller's "Super" which had already been in discussion with the wartime Manhattan Project {Rhodes 1986}.

Limited to a cryogenic nuclear weapon, the Mk-16 was 5 ft diameter, 25 ft long and 40,000 lbs with 6-8 MtE yield. It is the only deployed cryogenic nuclear device. The entire system of charging and managing deuterium boil off introduced a range of complexities. Even the massive B-36 required modifications to carry the weapons with only one aircraft configured. The proof of solid fueled thermonuclear capability quickly replaced this monster. None of these exist, but it can probably be credited as the lowest production nuclear weapon.

Mk-17 and Mk-24
Service 1953 - 1957
200 Mk-17s and 105 Mk-24s
1st storable thermonuclear weapon
Largest & heaviest US nuclear weapon

These were the first solid fueled lithium deuteride thermonuclear weapons, the largest weight and size bombs deployed by the US. The Mk-24 was tested in the Castle Yankee test shot, the second highest yield US nuclear test.

Tripping Over Engineering: Going Nuclear

Both of these bombs were 5 ft diameter, 25 ft long and 40,000 lbs. Like the Mk-16, they were only deployable with the B-36. Mk-17 was 15 MtE yield and Mk-24 10-15 MtE. Both weapons had an in flight insertion of the primary capsule to increase safety of deployment. With the massive yields, a 64 ft parachute was intended to provide some escape time for the aircraft. No matter how you look at these, they are simply colossal.

The Castle Bravo test shot of 1 March 1954 was the first lithium deuteride test device. It's yield was 15 MtE, 2.5 times the forecast yield, the highest yield US nuclear test. Some crew members received direct radiation injuries. The crew in the instrumentation bunker had thought they were at a safe range, but were forced to evacuate the bunker wrapped in sheets to reduce direct contact of fallout. Fallout ranged far farther than anticipated, reaching the Japanese fishing boat Lucky Dragon No.5 causing radiation sickness and a death. This and the appearance of Strontium 90 in New York milk was a warning to the public on the extent of fallout {Schlosser 2013}.

On 27 May 1957, a B-36 carrying a Mk-17 was inbound for landing at Kirtland AFB. Procedure required pulling the drop lock pins so that with an approach emergency, the aircraft could be lightened by 40,000 lbs, quickly. The man sent back to pull the pin lost his balance in turbulence and unfortunately grabbed whatever was in reach, the bomb release cable. No need to open the bomb bay doors, they departed with the bomb. The high explosive detonated on impact making a 25 ft crater and hucking an 800 lb fragment a half mile {Mahaffey 2014}. The fact that the primary was not installed until the aircraft was well enroute to target prevented a nuclear detonation.

Survivors located at [NMUSAF], [CAM], [NMNSH] & [SACSM].

Tripping Over Engineering: Going Nuclear

How Much Is Too Much

**Mk-41/B41
500 bombs built
Service 1963 - July 1976 (US Bicentennial)
Highest yield US weapon, only US 3 stage**

Happy 200th birthday America, we're decommissioning our largest warhead. At 25 MtE yield, the Mk-41 was the most powerful bomb ever in the US arsenal. It was the only US three-stage weapon. Depending on third stage casing, it was either a clean or dirty version. The engineering progress was obvious in that while it could only be carried by a B-47 or B-52, it was down to 10,670 lbs. At twelve ft long and four ft body diameter, the largest yield weapon was approaching the size and weight of Fat Man which was 0.084% of its yield. It was considered the highest yield to weight weapon ever built. While it could be used as a freefall and airburst detonation. I would suspect that the crews would far prefer to operate with parachute and delayed laydown detonation, especially with the ability to produce third degree burns at 32 miles with a 4 mile diameter fireball.

It is hard not to compare this to the USSR Tsar Bomba which is considered to be designed for 100 MtE and tested from a Bear bomber delivered test shot at 50 MtE, 30 October 1961. The Bear was modified and flew sans bomb bay doors with the weapon distended. At 60,000 lbs, this was never intended to be anything except bomber delivered. This test determined the futility of weapons over a specific size due to the fireball blowing out the top of the atmosphere with a significant loss of the yield.

B41 improvements were expected to be capable of 35 MtE and stripping to a warhead making it Titan II deliverable. Yield 1,700 times Little Boy and 1,200 times Fat Man with the idea that you can actually keep staging the fusion to whatever yield you want. But both sides also found some critical issues to uber tonnage weapons. They are only useful for very specific missions like crushing missile silos. For soft targets they wastefully disperse massive amounts energy off target. Casing at [NMUSAF].

Tripping Over Engineering: Going Nuclear

Mk-53/B53/W53
~340 bombs built
Service 1961 - 1987
Highest yield US missile warhead

After the B41 was retired, the Mk-53 was the largest US nuclear weapon at 9 MtE. Weighing 8,850 lbs, it was approximately the same outer dimensions as the B41. Even though decommissioned in 1987, fifty remained as part of a standby stockpile only being dismantled in 2011. The W53 warhead on the Titan II missiles was the highest yield US missile warhead. Warhead at [TMM].

This device was specifically designed as a bunker buster to crush nuclear command bunkers and missile silos. For the B53 bomb configuration, this meant three 48 ft descent chutes and a crushable nose cone for delayed laydown detonation. With this removed from service, the largest remaining are "small" 1.2 MtE warheads of the B83. [MOA] & [NATM] have casings. [NMNSH] has one in the open and one in loading position under a B-52.

Tripping Over Engineering: Going Nuclear

Palomares Casings

Albuquerque NM   [NMNSH]
17 January 1966
Broken Arrow

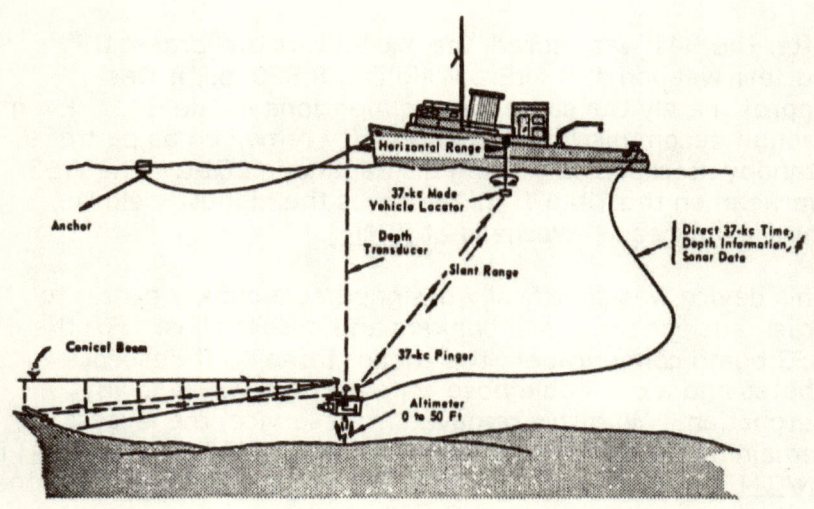

CURV II Recovery System {DTIC}.   ROV leading edge 1966.

After the military mitigation of a "Broken Arrow" (nuclear weapon accident), there is rarely much to examine beyond a large cleared area at the scene and a pile of debris and soil buried somewhere else.   Crew consequences range from reprehensible deaths to miraculous survival.   Thus is the nature of aircraft carrying lots of flammables, missiles with hypergolic fuels and radioactive devices packed with explosives.

On 17 January 1966 a B-52G and a KC-135 tanker collided during midair refueling during Chrome Dome continuous airborne alert operations.   Both aircraft and the four Mk-28 hydrogen bombs came down near Palomares Spain with only four survivors of eleven total crew.   Two of the weapons had chemical detonations, one struck the ground without detonation, one went into the Mediterranean.   The one at sea was recovered after an intensive search with a multiplicity of

Tripping Over Engineering: Going Nuclear

Navy vessels, including that little sub that everyone loves to love, Alvin. Extensive effort was spent on the clean up of radioactive materials from the two detonated weapons. Some clean up still continues.

But what of the two non detonated bombs? Both casings are in Albuquerque. It would seem that this mode of failure is about as good as one could expect. Parts of the thin shell nose and chute housing crushed and bent, thick warhead shell scarred and scratched, but all in all pretty much intact. There were issues from the accident, but it could have been far worse.

With gun type weapons there is a low fault tolerance once assembled. A single point triggering can provide a nuclear detonation. Even a significant impact can possibly drive the masses to a fizzle contact. With Little Boy, there was even a concern with an ejection over water moderating neutrons to the point of achieving criticality.

Implosion warheads, like the Mk-28, require a simultaneous firing of initiators to achieve nuclear detonation. Any error in this timing would prevent proper pit compression. It was fairly common to have an impact triggered chemical detonation with some dispersion of isotopes.

Over time, the nuclear powers developed improved safety systems. Sealed pits provided containment of the long life isotopes to reduce radioactive contamination from an accidental chemical detonation. Designs were moved to less sensitive explosives to prevent ignition from fire or shock. PAL systems were made more restrictive to fire. The systems became more self safing with incorrect input locking the circuits. Some contained manual safing that would fry the firing circuits to entirely prevent firing. This provides an immediate safety, but given time and know how, a warhead could be reconfigured for firing. Even with all of the weapons in circulation with transport, test firings and accidents during the cold war, there have been no accidental nuclear detonations.

On 24 January 1961, a B-52 broke up near Goldsboro NC with three fatalities from a crew of eight. In the break up, two

60

Mk-39 thermonuclear bombs were thrown from the aircraft. The 6,700 lb Mk-39 was a 3.8 MtE improvement of the 7,600 lb Mk-15, the first "lightweight" thermonuclear bomb. One bomb landed on parachute with three arming functions complete, including charging of the firing capacitor. Only the safe/arm switch in the off position preventing detonation. The second bomb impacted with extensive damage, partially armed, protected by one open switch. The disarming crew found the safe/arm switch in the armed position. Considered the closest potential accidental detonation. The damage and ground conditions of the impacted bomb lead to removal of the pit and burial of part of the physics package. The burial site with a 400 ft easement for farming only is easy to locate. While not from the accident, a Mk-39 casing remains at [NMUSAF].

Tripping Over Engineering: Going Nuclear

BOMBERS

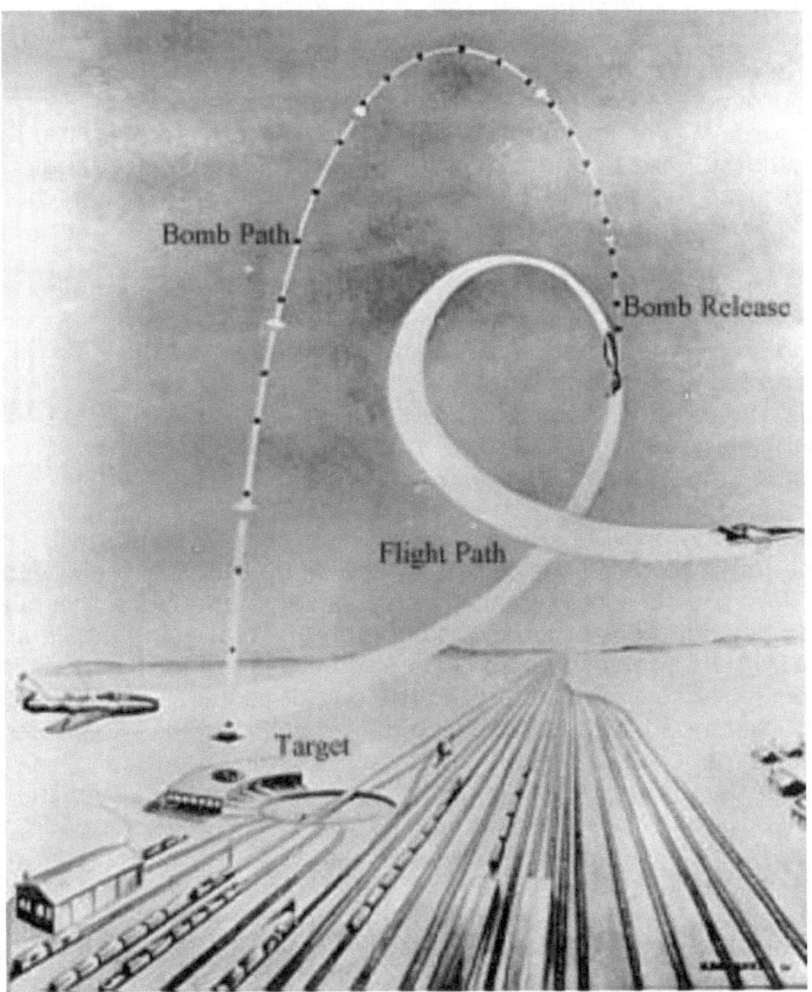

Artist's representation of a version of OTS bombing {NMUSAF}. Old school cool.

Tripping Over Engineering: Going Nuclear

At the beginning of the atomic era, big bombers were the only way to deliver big weapons. KtE yields required deployment from maximum altitude with target departure maneuvers to increase distance from the detonation and present the least affected aspect of the aircraft. The Army Air Force had extensive experience in high altitude precision bombing with the Norden bomb sight as the gold standard. Each variant improved the compensation of variables for increased accuracy. For its operational intent and technology limits, it would be hard to improve today. But, the key operational limit was the straight high level target run required to anchor all those variables.

The Navy concluded the war with direct naval engagement aircraft. Carrier aircraft were configured for close air support and ship engagement with torpedoes and skip or dive bombing. Most bombing systems required direct maneuvering with low altitude engagement. Even Navy patrol aircraft were typically equipped for low level engagement, torpedoes or depth charges.

In order to provide standoff from a target, particularly with the move toward lower altitude tactical operations, the speeds of jets provided ready energy for ballistic maneuvers. Hence the development of LABS toss bombing. While a typical pull up with 45° release using modern electronic flight directors is now fairly straight forward, in the day, it required specific approach path to a particular pitch up point. This was more problematic with mechanical computers and targeting systems. It was found to be far more precise to overfly the target and initiate pitch up with the LABS computer providing a pitch profile. This maneuver would be flown until the release trigger occurred at 110° pitch. This would toss the weapon up and back toward the target, bomber pulling through to inverted, rolling upright and running like hell. While technically called OTS for Over-The-Shoulder, it was more colloquially referred to as the "idiot's loop". In some cases the departure was more of a barrel roll with positive G maintained. While it was quite effective for small jets like the F-84, it was also done by the monster B-47 and puttering along in the Skyraider.

The Mk1 toss bombing computer was a gyro and mechanical

computer that fed inputs to the flight director. The pilot would set up on the target, engage the system, when the sight hit the target, pull up and follow the flight director. An F-100 held a 4G pitch up for two seconds before release. In order to keep the flight director needles centered, effectively flying on instruments through a ballistic maneuver, Les Frazier used an expressive metaphor, "easy as pushing an oyster into a slot machine" {Michel 2003}. The variable reliability of early mechanical autopilots made aviators reluctant to hand over control to the box at low altitude, high speed and all attitudes.

The yield of weapons mandated other than free fall air burst to protect the aircraft. Toss bombing gave some ordnance delay time for low altitude aircraft. High yield bombs received large parachutes to delay air burst detonation. Laydown delivery gave the aircraft more departure time as well as maximizing bunker busting with ground contact. Guided weapon standoff finally gave the paramount capability. Precision engagement of multiple simultaneous targets while permitting the aircraft to stay out of the hottest zone.

On Moscow Aviation Day 1955, ten Soviet Bisons were flown by the viewing stand multiple times to give the impression of more than 100 planes. The results of this bluff concluded in 1959 with the Air Force at their peak their bomber count with 1,366 B-47s, 488 B-52s, 174 RB-47s and over 1,000 KC-97 & KC-135 tankers in support {Polmar 1975}. The USSR ended the same period with ~150 long range bombers {Schlosser 2013}.

Once you can carry a nuclear bomb on an F-84, or even more so a Skyraider, all of your aircraft to some extent become nuclear capable. Pretty much all of the later fighter and attack aircraft, including the diminutive Skyhawk were designed for nuclear capability. Some Air Force fighters even had internal stowage capability. When you get to a warhead the size of Davy Crockett, there are no aircraft limits.

Tripping Over Engineering: Going Nuclear

AIR FORCE BOMBERS

While the US Air Force is now considered the ubiquitous air service, it technically didn't exist in WWII. It didn't even come into existence until 18 September 1947. Prior to that, it was always under Army control, although through WWII the US Army Air Forces had their own level of authority. When the Air Force was made independent, a significant portion was for the Strategic Air Command. This was exclusively for the issue of delivering nuclear weapons. Even the air defense radar and interception operations were primarily intended to be defense against nuclear bombers. Unlike the other services, the US Air Force's mission of birth WAS nuclear.

Tripping Over Engineering: Going Nuclear

B-29 Superfortress family

Superfortress - Super + Flying Fortress of predecessor, B-17

Since the beginning of warfare, therefore since the beginning of humanity, there has always been an effort to extend the reach of your weapons while protecting yourself from those of your enemies.

Since the beginning of aerial bombing, the push was for payload, range, speed and altitude. The B-29 was a leap ahead of the existing bombers. Compared to its contemporaries, even looking futuristic. Still an elegant beauty, less so for some of the progeny.

Roughly twice the weight of the other heavy bombers at the time, sometimes referred to as a super heavy. Although, when the B-36 was introduced, the B-29 suddenly became a medium bomber. It was similar to what the Germans had attempted with their WWII Amerika bomber program in order to reach the US. But the B-29 didn't just up the size and range, it was faster and higher. This technology jump did not come for free, B-29 development and production was the single most expensive US WWII weapons program. In excess of $3 billion, at least one billion over the cost of the Manhattan Project.

B-29s made two flights that were so significant that they dominate the memory of the aircraft. But, others in the B-29 family had their own contributions. Numerous B-29s and derivatives are on display. I have indicated noteworthy aircraft with their specifics.

Tripping Over Engineering: Going Nuclear

WWII
First flight 21 September 1942
Introduced 8 May 1944
3,970 Built
Most expensive US WWII weapon program

Of the innovations, there was one that had to be most loved by crews. A pressurized cabin finally permitted operation at altitude without supplemental oxygen and layers of heated clothing. The four defensive gun turrets were remotely operated through an analog computing system which provided assistance in lead targeting. The latest Norden bombsight provided precision in visual bombing. While the early R-3350s were maligned for reliability, every new higher performance engine had its range of issues. As a combat engine, there was expected attrition in the fleet. Over time, the R-3350 became a reliable and efficient military and commercial engine.

Operational experience in Japan found the speed and altitude of the B-29 as its best defense, leading to production of 311 B-29Bs and 65 Silverplate (specific to atomic bombing). The only defensive turret retained was the tail guns which received the AN/APG-15 radar fire control system. Of all defensive ordnance on bombers, the tail gun even stayed on the B-52 until 1991. The AN/APQ-7 radar bombing system was added for an obscured target capability.

B-29 operations started with Japanese targets in Thailand. The taking of the Mariana Islands provided range and runway capability to permit engagement of the Japanese mainland. Significant bombing started in February 1945 using different ordnance configurations. In an effort to bring the Japanese to surrender without the risks of an amphibious invasion, there was a firebombing effort similar to operations in Hamburg and Dresden.

Besides practically holding to the last man as was seen in Okinawa, the taking of Saipan had entire families, 22,000 civilians, that stepped off a cliff after Japanese propaganda had proselytized the cruelty of the Americans {Rhodes 1986}. It is easy to see where this was expected since their own

Tripping Over Engineering: Going Nuclear

army had forced as many as 200,000 Korean women to serve as military sex slaves and killing a million Chinese civilians with chemical and biological weapons {Schlosser 2013}. They were also informed on the extensive imprisonment by the US government against US citizens purely because of their Japanese descent.

Tinian island became the largest airport in the world. Bombing operations were performed with B-29s taking off every fifteen seconds for an hour and a half. Those that failed takeoff would burn next to the runway or go into the sea, as long as they didn't foul the runway, takeoffs would continue. A continuous stream of aircraft to the horizon. Several hours later, the stream would return {Rhodes 1986}. Early May 1945, Japan's four largest cities received 1,600 bombing sorties over ten days. Curtis LeMay stated, "we ran out of bombs. Literally." This burned out 32 square miles of city killing at least 150,000 {Rhodes 1986}. The actual deaths were probably multiples of the confirmed deaths. With millions of homeless and the limitations of emergency services at this level of devastation, unknown numbers simply disappeared.

Even with this devastation, lacking surrender, the bombing continued.

Leading up to Hiroshima, test dummies of Little Boy were dropped around Tinian to test targeting and trigger systems. Small groups of aircraft flew over Japan with some dropping "Pumpkins", Fat Man dummies with 6,300 lbs of high explosive. With the large raid traffic, these individual aircraft drew little response. When the weather was forecast to clear, the Hiroshima atomic bombing on 6 August 1945 by Enola Gay and Little Boy was almost textbook.

Even with this devastation, lacking surrender, conventional bombing continued.

Fat Man had an array of issues with its use. It arrived at Tinian with incorrect wiring. This required soldering fittings at a time when there was supposed to be no hot work around its 5,300 lbs of high explosive. A pump failure cursed 600 gallons of fuel to be destined to ride the roundtrip unusable.

Tripping Over Engineering: Going Nuclear

The fuse monitor started indicating that the weapon had armed itself. Rendezvous of instrument aircraft failed. The effectiveness of the previous firebombing missions obscured the primary target. After three runs, Kokura was abandoned for Nagasaki. The overcast sky forced bombing by radar when a hole in the clouds permitted momentary visual targeting. On 9 August 1945, after dropping Fat Man 1.5 miles off target, Bockscar diverted to Okinawa with limited fuel. By the end of the landing roll, two engines were dead from fuel starvation.

Even with this devastation, lacking surrender, conventional bombing resumed on Tokyo the following day.

Leslie Groves reported that another Fat Man bomb would arrive Tinian by 13 August 1945. It would be ready to drop depending on weather after 18 August 1945 {Rotter 2008}.

15 August 1945, the Japanese heard the voice of their emperor for the first time {Rotter 2008}. The message was in a formal Imperial Japanese which was not readily spoken among the population and surrender was not directly stated, creating confusion among the populace. Formal surrender was not signed until 2 September 1945.

Enola Gay
6 August 1945
Dropped 1st nuclear weapon on Hiroshima
Enola Gay, SN 44-86292 was the first aircraft to drop an atomic bomb in combat. On 6 August 1945 dropping "Little Boy" on Hiroshima. The second nuclear device ever detonated. Enola Gay has been restored and is on display at [NASMuh].

Bockscar
9 August 1945
Dropped 2nd nuclear weapon on Nagasaki
Bockscar, SN 44-27297 was the second and last aircraft to drop an atomic bomb in combat. On 9 August 1945 dropping "Fat Man" on Nagasaki. The third nuclear device ever detonated. Bockscar has been restored and is on display at [NMUSAF].

Enola Gay positioning to load Little Boy {AFHRA}.

Tripping Over Engineering: Going Nuclear

Post War
Until 1960

Post war, the B-29 was the only aircraft capable of carrying the nuclear weapons of the time. In May 1947, 101 B-29s flew over New York City in a "maximum mission effort" to demonstrate bomber capability. This was a third of the SAC bomber strength {Polmar 1975}. Through the Korean War, B-29s flew 29,000 sorties including the use of VB-3 and VB-13 guided bombs on vexing high value targets like bridges, continuing the move toward precision weapons. With introduction of the B-50, the B-29 continued in support roles. One of those support roles changed the reach of strategic bombing, the deployment of fighters and global logistics of the entire military, the tanker.

Tripping Over Engineering: Going Nuclear

FiFi & Doc

These two aircraft deserve mention for the significance of being the only two operating airworthy B-29s. Contact the guardians of these aircraft to find the airshows or flight days. It may even be worthwhile to find engine run maintenance days just to hear those 3350s. Both aircraft indicate that it is sometimes fortunate to not see completion of the final assignment.

Fifi was built by Boeing Renton as a B-29A then modified to trainer. Retired to desert storage then returned to service in 1953. Taken out of service in 1958 and assigned to China Lake as a missile target. Restored by The Commemorative Air Force and lovingly protected in Addison TX.

Doc was built in 1944 at Boeing Wichita. Operated as a radar calibration aircraft and later as a target tug. In 1956 taken out of service and also assigned to China Lake as a missile target. Acquired and restored at Boeing Wichita to flight status. Carefully attended by Doc's Friends in Wichita KS since 2013.

Tripping Over Engineering: Going Nuclear

B-50

1948 through 1965
370 aircraft built

Even with the advanced design of the B-29, it was adequately conservative to permit improvements. The wing and fuselage were kept relatively close with reinforcement in areas that were discovered to be marginal during extensive B-29 operations. But the R-4360 provided a ready horsepower increase and increased control surfaces managed that horsepower. This contributed increased range, speed and altitude along with 25% greater gross weight.

I have always thought the B-50 looked meaner and cleaner than the B-29. The sharp higher aspect ratio tail and the burlier profile of the R-4360. From the front, those big engines and chin intakes make it look like there is more engine than airplane. Had jets held off for a few more years, a turboprop upgrade could have made the next major performance jump.

Three complete aircraft are on display at [NMUSAF], [PASM] & [CAM].

Tripping Over Engineering: Going Nuclear

Lucky Lady II
2 March 1949
1st around the world non-stop flight

Lucky Lady II looped hose refueling {NMUSAF}. Notice the nacelle and tail difference between the B-29 and B-50.

The US had an emergency war plan in the spring of 1948 called HALFMOON. In the event of a Soviet invasion of Europe, the US military was expected to be outnumbered, perform a fighting retreat and within 15 days of first attack, retaliate with 133 nuclear bombs on seventy Soviet cities. They referred to it as "the nation killing concept". In order to provide the range with B-29s and B-50s, it would require some crews to make one way missions. Per Major General Earle E. Partridge, "It will be the cheapest thing we ever did. Expend the crew, expend the bomb, expend the airplane all at once. Kiss them goodbye and let them go." {Schlosser 2013}. Six years after the Doolittle Raid, there remained a concept of crews making their own way after bombing.

The entire focus of the B-36 plan was to provide a heavy bomb load with extensive unrefueled range. But, the fleet deployment delays of the B-36 left the B-29 and B-50 to do the heavy lifting. While there were experiments in aerial refueling back to the 1920s, the US had not progressed significantly since it was not considered beneficial for most combat operations.

Tripping Over Engineering: Going Nuclear

Strangely enough, through the 1930s, the progress in aerial refueling was not military, but commercial airline. FRL (Flight Refueling Limited) developed a looped hose method of fueling. In 1946-47 some airlines experimented with FRL systems on specific routes. A B-29 with an Air Force contingent landed at Heathrow airport to meet with FRL. After a few days discussion and payments, the Air Force bought their way into aerial refueling with two sets of looped hose refueling systems and an order for forty more.

In March 1948, the Air Force started modifying B-29s into tankers. In June 1948, all new B-50s were to be receiver configured with existing planes modified to single point fueling for easier conversion. The looped hose system required a receiver crewman to snag a cable and winch the hose to a hatch and fuel connection. December 1948 a B-50A using looped hose supplied by KB-29M delivered a dummy bomb load over 7,730 miles providing the range of a B-36.

Setting a record provides a public demonstration of capability to both your enemies and your electorate, especially if it is the first. On 26 January 1949 the Air Force commenced secret preparations for the first around the world nonstop flight. Five B-50As were prepared for the flight. One would launch and if it failed, off would go the next. At least three KB-29Ms were on standby at each of the four fueling locations. C-54s and C-97s with logistic and technical support crews were also standing by. Global Queen lost an engine and was on the ground after sixteen hours. Lucky Lady II with a crew of fourteen left Carswell AFB and apparently with no notable complications and eight dangle and grab refuelings landed back at Carswell on 2 March 1949 after 94 hours 1 minute with 23,452 miles in the air.

Refueling quickly moved to a probe and drogue system still used by the Navy. It provides a nice target basket so the receiver aircraft could fly into engagement. The trailing hose does not require manhandling and permits a higher flow system. The "Boeing Boom" was developed at the request of the Air Force. For the large aircraft in the Air Force fleet, this provided several benefits. The tanker flies straight and steady, the receiver flies into low trail position, the boom

Tripping Over Engineering: Going Nuclear

operator flies the boom to engagement. This allows fluidity of motion without forcing something the size of a B-52 to hit a two foot diameter target which may be flailing in turbulence. The boom permits a much higher flow rate which is important when you are feeding something the size of a B-52. Both of these systems also permit significantly higher operating speeds than looped hose. By the end of 1950 there were 126 KB-29Ms operating as tankers refueling bombers and fighters. Bombers could be launched heavy on ordnance and light on fuel, then topped in the air. Jet fighters with notoriously short range and limited navigation capability could be mothered by a tanker across any ocean. It changed the way aircraft performance was defined, maximum flying weight and maximum take off weight were no longer synonymous. It changed military aviation. The rest is history. {Smith 1998}

The fuselage of Luck Lady II is at Planes of Fame Chino CA.

Tripping Over Engineering: Going Nuclear

B-29 Family Nuclear Tests

Of the 1,026 US nuclear tests, only 52 were airdrops {DOE 2015}. Of those 52, 16 (almost a third) were by aircraft of the B-29 pedigree.

Crossroads Able, 1 July 1946, Dave's Dream (one of the original B-29 Silverplates and monitor to the Nagasaki drop) drops Gilda, a Mk-3 similar to Fat Man on Bikini Atoll. With extensive press coverage, science coverage and 95 target vessels, the first post war nuclear weapons test {DOE 2006}.

Ranger, 27 January - 6 February 1951, four Mk-4s and one Mk-6 dropped by B-50D bombers with airburst over Frenchman Flat [NNSS] {Sublette}. Interesting aside, the AEC's public information office for the Ranger series was in the El Cortez Hotel on Fremont Street in Las Vegas. For most people, one of the last old school casinos. For me, my father-in-law's favorite casino through the 80s and 90s.

Buster-Jangle, 28 October - 5 November 1951, four drops by B-50 bombers with airburst over Yucca Flat using three Mk-4s and a prototype Mk-7 [NNSS] {DOE 2006}.

Tumbler-Snapper, 1-22 April 1952, four drops by B-50 with airburst over Frenchman and Yucca Flats using three Mk-4s and a boosted Mk-7 [NNSS].

Upshot-Knothole, 6 April 1953, Mk-5 airburst by B-50 over Yucca Flat. 8 May 1953, Mk-6 airburst by B-50 over Frenchman Flat [NNSS] {DOE 2006}.

Besides the dropping of nuclear devices by the B-29 family, B-29s were always a core of airborne science monitoring of the tests from Trinity up into the 50s. Kirtland AFB had what became the 4925th Test Group that was tasked with aircraft nuclear qualification and aviation support of nuclear testing.

Tripping Over Engineering: Going Nuclear

Derivatives

It is hard to see the B-29 family without noticing its influence beyond carrying weapons to the target. These aircraft are in many locations, but two collections with multiple aircraft from the family are [PASM] & [NMUSAF].

B-29:
Wartime improvements with multiple factories prompted many minor variations between the aircraft, even within models.

Postwar main variants were tanker versions with looped hose, probe and drogue and boom systems. Sixteen were configured for air-sea rescue with an airdroppable lifeboat that clamped over the bomb bay. It was mothership for the X-1 on the first manned supersonic flight, also carrying a range of test aircraft including the parasitic fighter program and Tip-Tow tests. It pioneered airborne early warning systems and developed jet engines on extendable mounts from the bomb bay.

A B-29 was modified for test dropping the 44,000 lb T-12 bomb to free up B-36s for nuclear missions, probably stripped bare with minimum fuel. Long before satellites, there were weather and reconnaissance versions. Broadcast from the Republican National Convention on 23 June 1948 relayed television from Philadelphia to a nine state region via B-29 with the STRATOVISION communications relay.

XB-39:
The wartime XB-39 had Allison V-3420s as a test bed in the event the R-3350 was excessively problematic. The R-3350 was apparently good enough that the XB-39 was killed. While water cooling on these dense engines eliminated the issue of cooling air flow, pairing V-1710s through a gearbox would have its own range of issues. The German's also failed to solve the paired engine problem in the Heinkel He 177. The V-3420s were never used in an operational aircraft, while the R-3350 became a standard for efficiency in military and airline service until the era of the jet.

Tripping Over Engineering: Going Nuclear

B-50:
More power on the B-50 provided more performance for similar derivatives to the B-29 with tankers, reconnaissance, weather and radar. A tracked landing gear was tested. A tanker version received two podded J47 jets to assist in refueling the newer faster aircraft. A super B-50, the B-54 was considered with a larger single bomb bay, thirty percent gross weight increase and the need for outrigger wheels to protect the longer wing. The runway and taxiway modifications required for the use of the outriggers was a key justification to kill the program. But, strangely enough, two of the main strategic bombers in later years, the B-47 and B-52, both demanded outrigger gears.

C-97   1944-1978:
Towards the end of the war there was increasing need for air mobility. Take a B-29 fuselage, cut off the top half and mount a larger upper fuselage with relocated control surfaces devising a large capacity transport. It may look corpulent, but the XC-97 set an air transport record carrying 20,000 lbs of cargo from Seattle to DC at 383 mph. It was one of the early ramped drive on freighters with performance upgrades paralleling the B-50. It was particularly useful for SAC when they found that tanker operations also required supporting transport capability. The KC-97 provided ready support for the new missile weapons systems, but the limitation of the ramp entry would not provide a full ICBM transport capability until deployment of the C-133. Fifty six running R-4360s were built for airline service with Pan Am. The CIA performed signal snooping near West Berlin with three converted tankers.

With the arrival of the B-47, the KC-97 had to "toboggan" the refueling. The tanker would go to altitude, descend for adequate B-47 airspeed, engage and fuel during descent with the B-47 breaking away at a reasonable altitude and everyone having to claw back to altitude. The jet assisted KC-97s finally had adequate speed for level refueling with the jets. In the event of a bomber fuel emergency, the KC-97 could also provide its own operational avgas to the troubled jet. Once a jet is running, it will burn a range of fuel, potentially putting some wear and tear on pumps and hot section, but better than crashing a plane. I suspect that the Air Force preferred for an old KC-97 to gave up all of its fuel with the

Tripping Over Engineering: Going Nuclear

crew hitting the silk rather than losing a new expensive B-47 with a live nuke on board.

There are a range on display, but [MOA] has a KC-97L, the final and most advanced iteration.

Guppy:
Cheap retired C-97s and the need to move large aerospace parts led to the Guppy family by Aero Spacelines. These modifications provided a cavernous upper fuselage with the nose hinged to provide direct loading. The Pregnant Guppy hauled Gemini and Apollo parts from 1962 until 1979. The Mini Guppy carried large parts for aerospace companies from 1967 until 1995. A sister aircraft was built and upgraded with Allison 501 turboprops, but was lost during flight testing within three months of its first flight. Five Super Guppys were eventually built for use by NASA and aircraft builders with a larger cargo floor and turboprops. Modified wings and tail improved lift and control issues of pushing the big bubble through the sky. A NASA Super Guppy is at [PASM]. NASA continues to operate the last flying Super Guppy.

International:
The US refused to supply the B-29 to the Soviets in WWII under the Lend Lease Program. In 1944, B-29s on ferry flights made emergency landings in the USSR. The Soviets refused to return the aircraft and three were provided to Tupolev. The aircraft were different models from different plants so the Soviets could see the development path. In the first year of reverse engineering 105,000 drawings were made. Rather than copying the R-3350, they developed an engine based on their R-1820 production license. 847 Tupolev Tu-4s were eventually built. It dropped the first Soviet nuclear test bomb which was also built from extensive information on the Fat Man bomb. Similar to the US, it matured to tanker, patrol, reconnaissance and test aircraft carrier. There were prototypes of cargo, airliner and improved bomber with a super version along the lines of the B-54. Turboprop conversions were implemented, in some ways leading to the Tu-95 (for lack of a better term, the Soviet B-52). Turboprops provided to the Chinese which were tested as their first airborne early warning platform. The Soviets retired the Tu-4 in the 1960s, the Chinese operating some until 1988.

Tripping Over Engineering: Going Nuclear

B-36 Peacemaker

384 built
Service 1949 through 1959
1st global range nuclear strategic bomber
Uber bomber

Peacemaker - One who brings about peace, especially by reconciling adversaries

B-36 with XC-99 {NMUSAF}.

The concept of the B-36 originated with concerns that the US would have to engage Germany from a North American base. Progress in the Pacific Theater reinforced the need for a high capacity long range bomber. While work started during the war, first flight would be post war. Once the Cold War began in earnest, the size and capacity provided the most capability with the large early bombs. The early thermonuclear bombs were specifically built around the B-36. Mostly due to tonnage, the B-36 dropped bombs for five of the 52 air dropped nuclear tests. There were even special super heavy conventional weapons that were designed specifically for the B-36.

It has the longest wingspan of any combat aircraft. While through the late 1940s and early 1950s, there were numerous piston engined aircraft with jet boosting, adding four J47 jets on the B-36 always looked like a more natural conversion than on most other aircraft. The pusher configuration of the six R-4360s spinning huge 19 ft propellers on a massive slightly swept wing contributed to this appearance. These were the largest modern (no, the Linke Hofmann is not modern) propellers at the time, not to be eclipsed until later Soviet NK-12 engines. These could still be considered largest for modern piston. The twin pods resemble the B-47 inner

Tripping Over Engineering: Going Nuclear

mounts, also with J47s. The six turning four burning configuration made it made it the most copious engined aircraft since the Dornier Do X. The first prototype ran single wheel mains with the largest tires of the time at over 9 ft diameter. The production landing gear was four wheel bogies, after prototyping a tracked gear to reduce ground pressure.

The progress of jets and missiles would soon eclipse the capability of the B-36, but in its early years it had particular defensive advantages. The relatively low wing loading gave it altitude and maneuvering performance that was problematic even for jet interceptors at the time. In the days prior to aerial refueling, it had an combat range of 7,500 miles with 10,000 miles in ferry configuration. This made it the only aircraft able to directly reach Soviet targets from a US launch while the B-29/50 and B-47 required forward staging or inflight refueling.

A flight completed 12 March 1949 carried a 10,000 lb bomb load 5,000 miles to be dropped in the Gulf of Mexico then routing home for a total of 9,600 miles over 43 hours 37 minutes. On 29 January 1949, a B-36B flew from Carswell to Muroc with two 42,000 lb dummy bombs. The articles refer to "Grand Slams", most likely T-12s, the heaviest conventional bomb ever in US inventory. These would have had equivalent yield to some of the miniature nukes. One was dropped at 35,000 ft and the other at 40,000 ft, with the aircraft returned to Carswell. I hate to think how much of Texas was covered just reaching cruising altitude. It is pretty simple to drop a big huge bomb suspended from center of gravity. With two, you have to spread them out somehow. There would be a number of fuel and ordnance CG checks repeated en route to Muroc. Probably even dropping the first with a hand near the release for the second if there was a harsh pitch condition.

The NB-36H was a true "nuclear" bomber with an installed nuclear reactor. While the reactor was not directly configured for propulsion, it was operated in flight to test radiation exposure and crew shielding. The only other aircraft similarly flown was a Soviet Tu-95. While the nuclear bomber concept was considered to be achievable, the risks and costs were

Tripping Over Engineering: Going Nuclear

somehow considered excessive. The Soviets canceled theirs for the same reason.

It could have been the first bomber with long range standoff weapons if the Air Force had continued development using a B-36 as the carrier for the supersonic intercontinental 112,000 lb Boojum cruise missile. I would have bought a ticket to see that.

Its massive size permitted the concept of carrying parasite fighters for either defense or reconnaissance. Some were modified production fighters, but an interesting exercise was the XF-85 Goblin with thirty minutes fuel, four 50 calibers and no landing gear, the antithesis of a B-36. Ten B-36s were modified as recon aircraft to carry a GRF-84F recon fighter, possibly the strangest reconnaissance aircraft system operated.

As an operational precursor to the U-2, picture a special lightweight photo reconnaissance model reaching 58,000 ft altitude. It was built to defend itself and carry gigantic bombs over astounding distances. So strip the weapons, take off with full fuel, supercharge as much manifold pressure as possible. After hours of flight, higher and lighter, higher and lighter. Arriving on target, the wing loading is low from fuel burn. High and full throttle over the photo run, then a nice efficient cruise home.

Two were built entirely with jet propulsion as the YB-60. But the simple repowering of wartime piston engine design was not adequate. Weapons were getting smaller and lighter, the new B-47 and soon to arrive B-52 were functionally superior. A B-36 jet pod also drove an experimental locomotive setting an American rail speed record with the Black Beetle. Having found this propulsion less than economical, they converted it into a jet rail snow blower. Besides snow, it apparently also blew the ballast out from under the track. "If anything is worth doing, it's worth doing to excess", Edwin Land.

There is a strange naming relationship with the LGM-118 Peacekeeper ICBM. It was also originally intended to be called the Peacemaker. Apparently, between the 1940s and 1980s, the military was changing the concept of keeping the

Tripping Over Engineering: Going Nuclear

peace versus beating peace into an enemy with the biggest stick. I think that the world's largest bomber carrying the world's largest bomb is more a concept of suppressing adversaries than reconciling adversaries.

Like the last giant dinosaur, its species was rapidly reaching extinction, but it managed to delay its demise if only for being the biggest critter around. While it was far before my time, I did have an indirect connection with the B-36. My first boss at Cameron graduated Texas A&M in 1947. His first job had been at Convair Fort Worth, working on the tail skid for the B-36. Even a project of this size requires someone to work on each individual part.

Very early in the program, the B-36 was used as a base for a large cargo aircraft. The only XC-99 remains the largest piston engined transport ever built. It had payload and payload range records for its time. Since it was the largest in its time, the lone prototype ran 7,400 operational hours over 10 years delivering oversize heavy equipment, which was extensively B-36 engines and parts. It is currently in storage at [NMUSAF] pending funding for restoration. I saw the colossal XC-99 years ago, languishing next to the perimeter fence at Kelly AFB.

Of 384 built, only five remain. One each is at [PASM], [SACSM] & [CAM]. [NMUSAF] has a B-36F and an original single wheel landing gear along with a collection of components. An early prototype is owned by a private collector.

Tripping Over Engineering: Going Nuclear

B-47 Stratojet
Service June 1951 through 1969(bomber), 1977(recon)
2,032 airplanes built all variants
1st modern strategic jet bomber

Stratojet - Strato (Boeing prefix, 2nd layer of atmosphere) + jet (1st Boeing jet bomber).

B-47B RATO take off 15 April 1954 {NMUSAF}.

The B-45 Tornado was the first Air Force operational jet bomber. Jet engines provided distinct performance improvements, but the straight wing was still pedigreed to piston engines. Per Bill Gunston, the B-47 was "a design so advanced technically as to appear genuinely futuristic" {Polmar 1975}. The design of the B-47 allowed it to properly take advantage of jet capabilities with elements of design still seen on current military and commercial aircraft. It is the first MODERN strategic jet bomber.

Tripping Over Engineering: Going Nuclear

Wartime design had proceeded toward straight or slightly swept wing aircraft. Boeing technical staff accompanying the Army postwar inspection of German aeronautics facilities found swept wing test data and immediately wired orders home to change to the swept wing design.

The swept thin wings and early jet engines were a strain on the shorter airfields of the time. B-47s regularly used RATO bottles to assist take off from shorter fields and at high gross weights. RATO can't be tested before it is fired or killed after it starts. Takeoffs were initiated on jet with performance checkpoints on the roll. Hitting a particular checkpoint at a specific airspeed with all engine needles in the green confirmed performance. The takeoff would be abortable up to this point. The RATO bottles would be fired and takeoff was committed.

Lethargic throttle response and clean aerodynamics led to a unique double chute landing system. A small high speed approach chute was opened on final so that a higher more responsive throttle setting could be used. If the approach chute was cut on an aborted landing the next approach would be made under far twitchier conditions. This led to transition training with the crews repeatedly lugging the approach chute around the pattern with runway flyovers. On touchdown the main braking chute was opened.

The design concept at the time was that thin wings equaled high speed. Thin short wings work pretty good, thin long wings bring an entire array of issues. With the swept wing it becomes longer than its span with a range of torsional loads not seen on a straight wing. Where the wings were about the only engine mounting option for prior piston bombers, the compact jet engines allowed a range of options as seen on the Martin XB-51. The engine mounting on the B-47 solved more than just providing thrust. The forward pod configuration provided distributed flutter counterweighting using existing installed mass. With higher speeds are higher forces and the distribution of thrust and weight along the wing also provided more distributed stresses. At high speeds the ailerons did not directly alter the lift, but with limited torsional rigidity twisted the wing, a discovery termed aileron reversal. Vortex generators were developed to energize boundary layers which

Tripping Over Engineering: Going Nuclear

behaved differently in the transonic regime. Besides these lessons being applied to the B-52, they are seen on every jet airliner currently in operation.

The pilots still had wheels, but they sat in tandem in a bubble cockpit like a fighter. The crew spaces were tighter than other large bombers, but even the pilots could shinny out of the cockpit with a crawlspace on their left. The bombardier/navigator sat in the nose all alone. For the size and duration, the workload on three crew was intensive. Two spare crew could be seated in the crawlspace if necessary. An upgraded B-47 was initially considered to develop the new B-52, but LeMay insisted on a new design without the inherent restrictions of a derivative. Two of these were range and capacity. At the time, there were weapons that could only be delivered by the B-36. The B-52 needed the capacity to carry these massive devices. The B-47 also lacked the wanted intercontinental bombing range for many nuclear missions, requiring aerial refueling to round trip the sortie. The limited range led to the MB-47B concept as an unmanned drone, what could have been the world's largest cruise missile.

The substantial performance of the B-47 provided for extensive use in other applications. A tenure spent somewhere around or over every area of concern or combat. Before the U-2 was developed, the RB-47 was already over Soviet airspace on reconnaissance of all kinds; photo, listening to enemy radio traffic, tracking radar signals for anti-aircraft systems, relaying radio long before satellite communications. Flown as a test bed including; turboprops, the first Air Force fly-by-wire, aerial refueling and nose configurations for other aircraft. Data intensive configurations seated crews in bomb bay spaces and bristling with antennas. While the B-47 never dropped a bomb in combat or a nuclear test bomb, the number in service made it extensively the backbone of SAC from introduction until well after B-52 commissioning. Even with a large fleet to scavenge parts as a retirement plan, two issues were easier solved with newer aircraft. Time in the air was cockpit intensive and time on the ground was maintenance intensive, sacrifices made to be on the bleeding edge.

Tripping Over Engineering: Going Nuclear

At the time of B-50s and the B-36, the B-47 was the IT bomber with speed, altitude, range and capacity. On 8 February 1949, one of the two XB-47 prototypes broke ALL transcontinental speed records at 607.8 mph. With more long range speed records through 1954 {Polmar 2009}. Initially a typical high altitude level bomber, one would not consider an aircraft of this size as aerobatic, but in 1955 a B-47 at max combat weight of 130,000 lbs did an idiot's loop at 2.6G tossing an 8,850 lb dummy bomb vertically {Michel 2003}.

Jimmy Stewart, upon seeing the new B-47 while shooting "Strategic Air Command" said "She's the most beautiful thing I've ever seen in my life" {Schlosser 2013}. The bicycle gear had a compactness and position to appear flying while on the ground. While many got the flash protective white bellies, the remainder of the aircraft was typically bare aluminum. Every joint, every rivet is visible.

25 aircraft are on display. A pretty one is at [NMUSAF]. The one at [NMNSH] conveys what Jimmy Stewart saw.

B-52 Stratofortress
Service 1955-2050?
744 Built, 75 in service
THE BOMBER
Stratofortress - Strato (Boeing prefix) + fortress of prior Boeing bombers.

XB-52 with tandem cockpit and notable wing deflection {NMUSAF}.

So many records, so much capability, more combat service than any aircraft and scheduled to be in service longer than any other, the quintessential bomber. Contract in 1946, first flight on 15 April 1952 with the last new airframe rolling out on 26 October 1962. Operational to 2050 under current plan. With the B-52, the Air Force finally had the weight capacity and range approaching the B-36, the aerial refueling and speed of the B-47 with much better crew conditions and handling. 744 built of all models, 75 B-52Hs remain on active duty so you can still see them flying. A range of B-52s are on display, some notables are indicated below.

The B-52 learned a lot from the B-47. It had thicker wings which allowed fuel tankage and improved handling. This thicker wing better controlled torsional deflection, but it also added spoilers for improved roll control with reduced risk of aileron reversal. This was more improved in later models with a thicker section wing and turbofans. The thick section gave nicer flight characteristics and corrected wing fatigue issues. The B-52G finally replaced the classic aileron with

Tripping Over Engineering: Going Nuclear

spoilers and a small mid span feeler aileron, eliminating many of the torsional control loads.

The bicycle gear from the B-47 was greatly improved with a drift gage and gear drift alignment. Simply settle in on final with the gear down. Turn the drift gage until the line is on the path of movement and rotate the gear up to 20 degrees of drift correction. This gear design with mirrored retraction also made all four main gear identical.

The B-52 completed Operation Power Flite on 18 January 1957. This was the first jet nonstop around the world flight. It covered 24,325 miles in 45 hours 19 minutes even with having to fuel from the much slower KC-97s. One particular lesson learned from Power Flite was that something better than the KC-97 was needed for refueling a B-52 {Smith 1998}. It set an assortment of long distance speed records. On 11 January 1962 finally beating the 1946 Truculent Turtle unrefuelled distance record with 12,532 miles. With the exposure of large bomber bases, it was necessary to maintain an airborne alert of weapons. Using lots of fuel and lots of aerial tankers, a continuous airborne alert was maintained. To maintain six aircraft continuously airborne required an entire wing of 45 B-52s {Miller 2001}.

In the midst of the Cuban Missile Crisis, knowing full well that Soviets were monitoring Johnson Island tests, A B-52 released a test weapon on 27 October 1962. This was Operation Dominic and strangely enough this shot was named CALAMITY. (during the crisis there were two other tests Chama 18 Oct and Housatonic 30 Oct. These were all B-52 airdrops. Housatonic was the last US airdropped test) {Dobbs 2008}. The B-52B at [NMNSH] flew on Dominic and Redwing. It is the last remaining B-52 to have dropped a nuclear weapon and the first to have airdropped a thermonuclear weapon on the Cherokee shot. B-52s performed a total of 30 test bomb drops, making 58% of US nuclear airdrops {Polmar 2009}. The [NMNSH] aircraft is sitting with a B53 in the loading position and a Hound Dog on its loader in the flight position.

While it looks big and slow now, it was an effective high altitude nuclear bomber. There was a move toward standoff weapons to better penetrate air defenses with the Hound Dog. Technology in anti-aircraft weapons pushed the B-52 down to the ground like other bombers. Imaging and navigation systems permitted all weather low level operations. It was probably far more effective than one would expect for a 220 ton plane flying with its feet in the trees. Its real forte was discovered in Vietnam and beyond. The big bomb bays used for huge early nuclear weapons were quite useful for large conventional bomb loads. NASA found this useful on the two used to launch the X-15. One of these is at [PASM]. Balls 8 also launched an extensive range of NASA air launched devices, it is currently at the North Gate of Edwards AFB.

The B-52 entered service with one item dating from a WWII, the tail gun. There was only one enlisted crew on a B-52, firing a quad fifty on the A-G models and a 20mm gatling on the H model. This was true all the way up through Desert Storm. Then on 1 October 1991, the tail guns and the tail gunners were no more. The gun may not have seen much action but there had been hotter times. During Linebacker II on 18 December 1972, two Mig 21s approached a B-52D. As one approached, SSgt Samuel Turner fired a six second burst that blew apart the Mig. The other Mig turned away on reconsideration. Five B-52 gun kills were claimed during Vietnam, only two were confirmed, Turner's was the first {Grier 2012}. This aircraft is at Fairchild AFB.

Official code name, "Senior Surprise", code name itself classified, with crews calling it "Secret Squirrel". At the start of Desert Storm on 16 January 1991, seven B-52Gs lumbered down 9,000 ft of runway heaving 244 ton aircraft on their mission. 14-1/2 hours later, they they were dropping an extremely high security precision weapon on high value targets. The AGM-86C started life as a nuclear cruise missile with terrain radar guidance. A conventional warhead replaced nuclear and radar guidance replaced by a new high tech radio positioning system, GPS. With one aircraft running on seven of eight engines, they fired 35 of 39 missiles due to software issues. This left the aircraft to slog home with extra drag and greater than forecast headwinds. But with heavy aerial refuelling they rolled onto Barksdale after 35 hours in the air and over 14,000 miles. This shattered the record for longest range bombing sortie of 6,600 miles by British Vulcans

Tripping Over Engineering: Going Nuclear

to the Falklands {Tirpak 1994}. While the mission provided a particular benefit in destroying specific command and control targets, some claim that it was a publicity stunt to demonstrate that the Air Force could reach any target in the world by air. No, it wasn't a stunt, the Air Force can reach any target in the world by air. One of the Silent Squirrel aircraft is at [PASM].

While no one discusses the B-52 as a nuclear bomber anymore, it initially hauled heavy tonnage weapons and it is being upgraded for more tonnage weapons. With a 70,000 lb ordnance capacity, the two wing pylons are limited to 5,000 lb each. Proposed upgrades will permit 20,000 lb each, allowing a MOAB under each wing. MOAB has an 11 tE yield, similar to some prior Army battle field nukes.

Even with its age, it has a lower operating cost than many current aircraft. So much in technology is about replacements and the B-52 keeps outliving its replacements to the extent that a ninety year service life is considered by the Air Force. I met two generations that have flown the B-52, there is a three generation family of B-52 pilots. The last new one was built before I was born and I will probably be dead before it is retired.

Tripping Over Engineering: Going Nuclear

F-84 Thunderjet
Service 1947 - 1960s, foreign to 1972
7,524 airplanes built all models
1st fighter configured to carry nuclear weapons

Thunderjet - Thunder prefix used by Republic Aviation and, well, it's a jet.

First built XF-84, gorgeous before all the pylons are bolted on {NMUSAF}.

Six fifty caliber Brownings (same as a P-51), a straight wing and the first US production axial flow jet. Typical of early jets, poor take off thrust, screaming laboring lumbering until deciding to give up the earth. But once airborne, impressive enough to become the first plane of the Thunderbirds.

It tracked well, slogging up to 4,450 lbs of ordnance into the air under the right conditions. If necessary, a couple of RATO bottles could help heave it into the blue. While limited as an air superiority fighter, it was productive in delivering tonnage over Korea. This merged well with the nuclear weapons technology achieving the Mk-7. Hence, the F-84G was the first fighter configured for delivering a nuclear weapon. The speed and aerobatic capability were a great benefit in

Tripping Over Engineering: Going Nuclear

delivering weapons via the idiot's loop {Michel 2003}.

There were several other items that made it even more significant as a nuclear bomber. It was the first production fighter with aerial refueling. This allowed mass long range movements of jet fighters. In 1953, seventeen F-84Gs flew 4,485 miles transatlantic for the longest nonstop fighter flight of the time. Considering that the Mk-7 had more yield than the first three models of nuclear bombs, tactical firepower was greater than the original strategic firepower.

The F-84 was extensively used in the testing of parasite fighters for the B-36 both as tip-tow (airborne connection of wing tips) and trapeze (hooking a frame under the fuselage). This went along with a the dream of heavy bombers always having their own personal escort. But, for the effort and risks of mating a fighter, a lot of other equipment could be added, particularly in the era of improved electronic detection and guidance.

In 1955, it showed its independence from such fussy items as runways, as the first manned aircraft to launch with a zero length launcher. Using the equipment at hand; take a Matador missile launch trailer, bolt the launch lugs onto the aircraft, load it lightly and bolt a Matador boost rocket on the belly. Voila, no runway, launch a jet fighter out of a tennis court. A concept forgotten with improved missile and aircraft capability.

The F-84 was the last step to ubiquitous nuclear bombing capability. As fighters continued to get larger and faster, nuclear weapons got smaller and more aerodynamic. Most following Air Force fighters with a bomber capability were designed inherently with the nuclear delivery capability. For all of its original limitations, the F-84 was the little engine that could. Numerous F-84s are on display, very nice ones at [MOA] & [NMUSAF]. [PASM] has straight wing and later swept wing models.

Tripping Over Engineering: Going Nuclear

B-58 Hustler
Service 1960-1970
116 Produced
1st supersonic strategic bomber
Hustler - One who deceives less skillful players to play for money.   Alternate, a prostitute.

RB-58A (58-1011) with single component pod mounted and two component pod being assembled {NMUSAF}.

In the effort for faster higher bombers, the B-58 was the first direct effort at a supersonic project. It was developed in the middle of the delta wing boom with many countries developing subsonic and supersonic aircraft. The delta provided large wing surfaces with thicker sections for structure. It also had a much longer effective chord with a greater center of lift movement transonically. All control surfaces aft prevent the use of flaps for lift augmentation. The low aspect ratio incurs high angles of attack and forces low speed flight on the backside of the power curve.

With Convair's extensive experience, why not build a great big fighter. While it would appear that history would dictate internal engines, four J79s were podded along the wing. This would greatly improve maintenance and survivability on catastrophic engine failure while saving a lot of intake and nozzle plumbing. Four big engines loud and proud.

Being designed for long range supersonic operation, it required advanced structural thermodynamic design. It used significant aluminum honeycomb and Mag-Thor alloys

Tripping Over Engineering: Going Nuclear

provided high temperature capability if somewhat radioactive. The speeds and altitudes also made crew ejection so problematic that each of the crew sat in an escape pod with an encapsulating clamshell. The crew were entirely partitioned from each other with no room to move or stretch, connected only by intercom. With top opening canopies the control and instrumentation entirely wrapped around all three onboard. The back seaters were not copilots, but managed navigation, bombardment and defense with only small windows looking out the sides of the cockpits.

Being the highest performance bomber, it also received the latest in electronics. Combined celestial and inertial navigation was passive and independent with the ability to correlate systems. The heavy radar use indicates a pre-stealth attitude, but all of that radar provided an order of magnitude improvement in obscured target bombing accuracy. Even with the speed capability, there remained the possibility of a fighter closing from behind, with a radar directed 20mm tail gun to welcome them. The modern psychological studies had found that young men were more responsive to a female voice, so in the super modern B-58, the voice of Joan Elms (actress and singer) would give twenty critical verbal warnings from a magnetic tape system. These admonishments were, depending on conditions, referred to as "Sexy Sally" and sometimes "Bitchin' Betty".

The Hustler had numerous records, particularly speed over long distances. A record of 8,028 miles supersonic was achieved even with a portion of the flight subsonic. John Denver kept the peace with his music and his father kept the peace with the B-58 setting several records. Besides the records, there were wondrous performance capabilities achievable in various regimes of flight. With light loading, it could climb at 46,000 ft/min and achieve altitudes of 85,000 ft. {Higham 1975}

A significant design feature for the nuclear mission was "the pod". With the size of early nuclear devices and the fuel consumption of hooking it at Mach 2, it was determined that a very clean slipper tank 5 ft diameter and 75 ft long would solve a lot of problems. The pod full of fuel with a W39 warhead weighed 36,000 lbs. On a nuclear mission, burn the

Tripping Over Engineering: Going Nuclear

pod dry, drop it on target and you were as clean and light as you ever would be. It gave the B-58 the best target departure capability of any bomber. Later development had a split pod so the fuel tank could be dropped with a smaller weapon pod carried to target. The development of smaller weapons allowed the addition of four Mk-43s under the aft fuselage for up to five target engagement. The pod permitted a relatively easy conversion with a photo reconnaissance package. A pod configuration was also considered with a rocket to provide a standoff weapon.

While the B-58 was adaptable to low altitude penetration, it was not readily adaptable to conventional missions. With a cost of ownership three times that of a B-52, it was destined for early retirement. Looking solidly built with the pod mounted, lean and long legged without it. Eight are on display: [NMUSAF] set three speed records, [SACSM] set a record for the longest supersonic flight, [PASM] was the last B-58 delivered. [NMNSH] has a two component pod with close accessibility.

XB-70 Valkyrie
First Flight 21 September 1964, Retired 1969
AV-1 is at [NMUSAF]
Too many superlatives to list
Valkyrie - (Norse) One of a host of maidens who choose who lives or dies in battle.

XB-70 take off {NASA}.

An Air Force bomber program canceled in 1961 with retirement in 1969 questions the logic of the math. Two aircraft in process were continued as high speed research aircraft. There were a lot of experimental aircraft at this time, but NONE were the size and performance of these two. The Valkyrie project started in the era of the higher, faster, bigger bombers. The concept was pursued until 1961 when they were eclipsed by ballistic missiles and anti-aircraft capability. Bomber development shifted to nap of the earth radar penetration. Much like the SR-71, the B-70 could not be an efficient and effective low altitude aircraft. Nor did its inherent design permit a range of conventional bombing operations.

As a supersonic research aircraft, it provided an instrumentation capacity previously unavailable, also making a massive size for handling. 194 ft overall length with only a

Tripping Over Engineering: Going Nuclear

46 ft wheel base the pilots sight line sitting 22 ft off the ground and cantilevered 64 ft forward of the nose gear with a big six engine cluster behind the mains. Everything was built to support supersonic flight, big canard, drooping wing tips for both supersonic lift and increased yaw surface, huge inlet diffuser for a rack of six jets, funny finger elevons, three big braking chutes. This made landing and takeoff an inconvenience to be dealt with. Final weight, 550,000 lbs, the worlds heaviest aircraft at the time. Flown to 74,000 ft and Mach 3.08. Heavier than a B-52, faster than a B-58.

Typical of a unique aircraft, it had functional peculiarities. On 30 April 1966 AV-2 had a landing gear short that partially retracted the nose gear cutting the tires and preventing extension. This happened on takeoff so the crew had time to work the problem with the engineers. They determined that the problem was a circuit breaker that would allow extension if shorted. Since aircraft electrical systems are not configured to permit the option of selectively shorting out circuits in the cockpit, one of the pilots dug a paper clip out of his flight bag and found a paper clip. He gained enough access to jam it into the wiring. Gear down and three green. But, the process had locked three of the four main gear brakes. And that is how they set it down, 196 mph with flaming shredding tires. Remarkably, the damage was so minor to permit return to flight in a fortnight. It was heavy and fast, which brings particular stresses on landing gears. Especially, when they are made as small and unobtrusive as possible in flight. This resulted in a tedious range of gear motions for retraction. On the 37th flight of AV-1, the main gear did not provide full extension. This resulted in a tip toe landing with the bogeys tripped up, sitting on four tires instead of eight. It was repaired and soldiered on. The AV-2 was hit by an F-104 caught in the wingtip vortex and crashed 8 June 1966.

Too late in development, too expensive to build, too expensive to operate. Neither a modification nor a derivative of an existing aircraft. No progeny of any kind, absolutely and singularly unique. A Cold War icon. It shows what engineering can accomplish even when the question is why.

Tripping Over Engineering: Going Nuclear

NAVY BOMBERS

In order to discuss Naval Aviation for the purposes of nuclear bombing, there is only one thing to discuss, carrier operations. If you are going to operate land based, then the requirement would be no different from the Air Force. When maximum runway length is 1,123 feet and dynamic, it brings about a very distinctive set of problems. Sure, you have energy management with cats and traps, but even that brings an additional range of design issues.

While the only strategic Navy nuclear weapons typically discussed are SLBMs on boomers, the first consideration was bombers on carriers. The same big heavy bomb as the Air Force with that short continuously moving runway.

B-25 Mitchell

Doolittle Raid
18 April 1942
1st US airstrike on Japanese homeland
16 plane carrier attack on Tokyo

Mitchell - The only military aircraft named for an individual. Billy Mitchell promoted air power in the Army and pushed for cooperation with the Navy in the building of carriers. Court-martialed for statements against Army and Navy commands for aviation related incompetence. In WWII (after his death) was recognized for his foresight and considered the father of modern air power.

B-25 taking off USS Hornet for the Tokyo Raid {NMUSAF}.

A plan rushed into operation, slightly over four months after Pearl Harbor, only for the purpose of striking at the hearts and minds of the enemy. If the enemy hits your country even if it is on the frontier, respond by hitting their capital and they will at least question the wisdom of what they have done, particularly if the populace has been propagandized as

Tripping Over Engineering: Going Nuclear

superior and untouchable. Using the far ranging legs of US power, the carrier, and the far reaching arms, the bomber, you can hit where you want, if you are resourceful enough.

The B-25 was selected for launch off of the USS Hornet to attack Tokyo and three other cities to a lesser extent. The B-25 was a relatively new Army medium land bomber with the balance of size, capacity and take off characteristics to achieve the goal, but range limits impeded a round trip. After bombing, the crews were to head toward an allied, or at least non-combatant country, if they could.

Since the attack was considered to a be complete surprise, the aircraft were configured to be mostly offensive. The only gun mounts retained were the upper turret and the nose gun. With the low altitude plan, these were expected to give adequate protection with some offensive capability. With the tail guns were removed, it was considered adequately exposed to at least want the threat of defense. Two black painted broom handles were placed in the tail gun position of each aircraft, some crews reported this as a effective fighter deterrent. The "Mark Twain" bomb sight replaced the Norden for the lower altitude bombing, looking like a V-notch gun sight on a protractor with a table of speed and altitude used to set the declination angle. The sight radius gave a bit of alignment for approach and bombs were released when the target crossed the sight point. A replica of this sight is at [AFAM]. There was a reduction of one crewman compared to a standard B-25 and the fuel capacity was almost doubled. Each aircraft carried three 500 lb high explosive bombs and a 500 lb incendiary cluster. This one way launch operation would set the tone for similar plans during the Cold War.

The balance of B-25 characteristics permitted it to perform a majority of aviation combat functions. Under a later Navy designation the PBJ-1 was extensively used by the Marines as an all purpose aircraft, using a range of bomb, gun and rocket configurations. But the carrier operations were not forgotten and on 15 November 1944 a modified PBJ-1 performed a full trap and cat launch on the USS Shangri La. Catapults were used more extensively toward the end of the war to allow for more aircraft on deck and more wind variability tolerance. Army aircraft were packed on transport carriers going to the

Tripping Over Engineering: Going Nuclear

Pacific. The simplest unloading was catapulting a lightly fueled and unarmed aircraft to hop to the local runway. Land, do a combat loading and you are ready to fight. This lead to Army aircraft configured to launch from carriers, with catapult fittings on P-38s, P-39s, P-40s, even P-61s. All late war Pacific bound P-51s and P-47s were cat configured from the factory {Friedman 1983}.

Another point of engineering was Jimmy Doolittle who planned and led the raid. A flight instructor at the end of WWI. He received what is considered to be the first US doctorate in aeronautical engineering from MIT in 1925. Following this as a racing pilot, he won his share of races and set several records, including surviving a win with the horrendous Gee Bee R-1. He set the stage for higher horsepower piston engines by influencing Shell to develop 100 octane gasoline. In 1929 he made the first flight from take off to landing entirely on instruments. While the Doolittle raid was high risk, it was never intended to be a suicide mission. Of 80 total crew, 69 eventually made it home. Post war, Doolittle continued to be called for Air Force and intelligence agency technical support, often to provide analytical ability to particular problems. He was chairman of NACA in the early space race, being offered and refusing the position of the first administrator of NASA.

There were 9,816 B-25s built and operated from early WWII through the 1960s in the US with foreign service to 1979. There are a large range of Mitchells on display in various conditions and several still flying the airshow circuit. I give reference to the Eglin [AFAM] and Pensacola [NNAM] aircraft if only for the relationship of these bases to the training and preparation for the raid.

Tripping Over Engineering: Going Nuclear

P2-V Neptune
Service 1946 to 1978
1,184 aircraft built
Neptune - (Roman) God of the sea.

During WWII, the Navy had used a variety of aircraft for long range patrol work. From specific Navy land and seaplane designs to converted Army heavy bombers, each with its own limits on capability. As the war moved along, the instrumentation and ordnance capability increased with the need for range. The flight time accumulation with long range patrols had specific impacts on maintenance requirements. While developed as a wartime design, the production of the Neptune did not start until post war. It arrived as a heavy twin radial engined aircraft. Wartime experience had significantly debugged the R-3350 engines to reliably and efficiently provide high horsepower. Later models would add a pair of J34s for improved take off and safer low level engine redundancy.

By retirement, they were also used by the Army, Air Force, CIA and multiple countries. It had used naval acoustic equipment in the jungle, picked up spies in the arctic without landing and armed as a good old gunship. After its military service, the weight capacity and robust airframe worked on as a forest fire bomber. While the two following sections bring the Neptune into the world of strategic bombing, it retained a Naval nuclear weapons capability from 1952 until the end of its active service. It was configured to carry the Betty and Lulu (one at [NMNSH]) nuclear depth charges. Both around 1,200 lbs and 10 KtE, probably named for ex-girlfriends. Because of the extensive service, there are a range of them on display with an array of provenances.

Tripping Over Engineering: Going Nuclear

Truculent Turtle
1 October 1946
Unrefueled distance record standing sixteen years
NAS Pensacola [NNAM]

Truculent - Eager or quick to argue or fight.

Truculent Turtle airborne before record setting flight {NHHC}.

Post war, there was significant animosity between the services as they competed for the ever dwindling military funds of peacetime. The Air Force B-29 was a proven long range aircraft having completed a 7,500 mile flight with rumor of a 9,000 mile attempt, the Navy felt the need to demonstrate their global reach in aviation. A study on the P2V range arrived at a maximum of 12,000 miles which closely approximated Perth Australia to Columbus OH. This was the holy grail of military range, halfway around the world, the ability to launch and reach any point on earth, by definition global.

The Turtle was one of fourteen aircraft with extensive range modifications provided by Lockheed. These were stripped of armament and turrets with fuel tankage installed everywhere there was space. To give the flight some sort of purpose,

Tripping Over Engineering: Going Nuclear

they were carrying one passenger, a nine month old gray kangaroo which was a gift to the National Zoo from the Aussies. They had a long range flight crew of four instead of the combat crew of thirteen. The additional fuel tankage was 5,000 gallons over standard for a total of 8,525 gallons. A fuel load of 54,560 lbs was well over half of takeoff weight, fuel alone over 85% of the early model design gross weight. For emergency purposes, the fuel dump system could drop 800 gallons in 20 seconds and 5,200 gallons in six minutes. Even standard oxygen, heating, deicing and some radio was removed. The engines were upgraded for higher cruise horsepower. The normal oil capacity of 180 gallons also had an additional 370 gallons, oil consumption and loss being the nature of radials. This lead to a total take off weight of 85,575 lbs. This is for an aircraft that typically had a max gross of ~64,000 lbs. The heaviest previous testing had been 20,000 lbs lighter. The tires were overinflated to carry the load, the struts bottoming out, there was a fear to turn the aircraft loaded on concern that it would wreck the gear. So, final fueling and RATO mounting was made after they lined up on the runway.

In the cool of the afternoon on 29 September 1946, they ran the throttles wide open and at 100 mph fired the RATO bottles to have any hope of getting off the ground. In the years before super runways were common for jets, this only gave 6,000 ft of runway to lumber down. When it lifted off, it became the heaviest twin engine aircraft to fly. It is still the heaviest twin engined piston aircraft.

The terrain allowed them to make a descending run out after clearing the trees. Slowly nursing up the altitude and turning back over Australia. One benefit of the planned route was that the overloaded start crossed Australia and the fuel depleted end was over the continental US. Proceeding with four hour watch rotations, they flew on with minimal radio contact and shooting stars for navigation. When they made radio contact in the US, it was the first in twenty hours. Hitting the US, there were concerns with the weather forecast, pressing on, they lived with turbulence and blue-white arcs of St. Elmo's fire off the props. On the radio, they would hear the static build and quiet with discharge. At one point they were at 13,000 ft on the remaining portable oxygen bottles,

Tripping Over Engineering: Going Nuclear

watching ice build up on the wings. But at 13:28 1 October 1946, they set down 11,236 miles 55 hours and 17 minutes after take off. The non-stop unrefueled record would stand until 1962 to be beaten by an Air Force B-52 at 12,532 miles. For piston/propeller aircraft, it would stand until Rutan's Voyager in 1986 went around the world unrefueled nonstop 26,366 miles.

Like so many record setting efforts of its time, it had little to do with showing the enemy your capabilities, but a vanity effort between the US services. There had to have been some interesting discussions with Lockheed, what if I was 10,000 lb over max take off weight, what about another 10,000 lb over that? {Gulliver 2011}

Carrier operations
28 April 1948
1st carrier born nuclear strategic bomber.

P2V-3C launch USS Midway (CVB-41) 7 April 1949 just after RATO bottle light off {NHHC}.

The P2V-3C was the first Navy carrier nuclear strategic bomber. How could you have never heard of the Navy strategic bomber program? Well, once upon a time 1946, the Navy contracted for a strategic carrier bomber that could carry the massive bombs of the time. In many ways, this was an effort to keep up with the Army strategic bombing capability. Unfortunately, the AJ Savage would not be available until 1950. So the Navy decided to proceed with an aircraft already in service which had demonstrated significant weight and range capability in the Truculent Turtle. Twelve aircraft were stripped of defense and patrol equipment, additional fuel added, bomb bay configured for a single atomic bomb, four RATO points per side, a belly ski for ditching, and a TAILHOOK.

Even with the modifications, these aircraft did not have the space to carry the preferred Fat Man, so Little Boy was reinstated for a limited production. With the catapults of the

Tripping Over Engineering: Going Nuclear

time, flying speed for a 70,000 lb gross aircraft was out of the question, solved using the entire carrier deck, full throttle into the wind and firing RATO.   Thundering smoking rattling down the deck with ten lavish feet of wingtip clearance from the island. Shipboard trials included take offs and touch and goes but never a full trap. After 128 shore traps, including one with a full reverse thrust on engines before touch down and one which shot rivets off the aircraft, it was decided that they would hoist the aircraft onboard and operate as a one way flight in the manner of the Doolittle Raid. The ditching ski was very much in line with the idea of a one use aircraft {Miller 2001}.

The first two aircraft rocketed off of the Coral Sea on 28 April 1948. A simulated attack was launched from the USS Franklin D Roosevelt off Jacksonville FL. Flying over the Bahamas, Panama Canal and Mexico to land in San Francisco 5,060 miles after 25 hours and 59 minutes, the longest range unrefueled flight from a carrier. To show the confidence of the Navy in the flight capability, following an airshow, USS Midway launched a Neptune with passengers returning to Washington National. Onboard were Secretary of Defense Louis Johnson, Chairman of Joint Chiefs Gen Omar Bradley, Secretary of Air Force Stuart Symington and William Randolph Hearst. It's impossible to picture having passengers of that status on what is more or less an experimental flight operation nowadays.   {USNI 2011} {Friedman 1983}

Once again, discussions with Lockheed. Hey, can we carry a 9,000 lb ordnance? Sure if you run it light enough. How about installing a tailhook? Installing a what? The program aircraft are gone, but numerous Neptunes are on display to draw an impression on the magnitude of the accomplishment.

Tripping Over Engineering: Going Nuclear

AJ Savage
Service 1950-1960
1st designed carrier nuclear strategic bomber
143 airplanes built
Only survivor at [NNAM]
Savage - fierce, violent, uncontrolled or primitive, uncivilized

6 December 1955, AJ traps aboard USS Yorktown (CVA-10) {NHHC}.

The AJ concept commenced in 1945 as a carrier based bomber with 10,000 lb ordnance capacity. During development, the Navy added the requirement to carry the Mk-4 nuclear bomb. This was an improved version of the original Fat Man bomb, but no smaller at 11,000 lbs, 5 ft diameter, 11 ft long.

The Savage was designed with two R-2800 radials and one J33 jet in the tail. There were a number of large aircraft with pylon mounted jets as performance additions. Only the Savage, Fireball and Mercator were operational aircraft originally designed with supplemental jets built into the aircraft. Of these, the Savage had much higher production and more extensive use than the other two.

The 1957 data sheet shows capabilities with lighter Mk-5 and Mk-15 bombs to a maximum range of 2,500 miles with a combat cruise speed of a howling 268 mph and a placarded red line of 483 mph. Pointed at the ground with its hair on

Tripping Over Engineering: Going Nuclear

fire, it could not outrun any of the fighters at the time. Since all sacrifices were made to heave a six ton weapon off a carrier, no defensive weapons were installed. The massive deck space occupied and hand pumping of wing retraction/extension did not endear the aircraft to carrier crews. Upon entering service, it was the heaviest carrier based aircraft, with its large weight (54,000 lbs max TO) and size (75 ft span) limiting it to a few ships. This was still in the era of straight deck carriers with hydraulic catapults.

The entire aircraft is about a single mission design in a period of extremely transient technology. Air Force and Navy dissonance for funding lead this to be one of two Navy aircraft to be labeled "Strategic Bombers". But, its large internal space and weight capacity permitted development as an early carrier based tanker. Savages refueled John Glenn's Crusader on his 1957 transcontinental speed record flight. But, like the same problems in the Air Force, a prop drive tanker running wide open with limited altitude capability does not provide a good operational window in supplying high speed and altitude jet aircraft. It was the last, for lack of a better term, "fast" carrier piston aircraft. All following fast combat aircraft would be jets.

Of the 143 AJs built, some reconnaissance models did photographic work for other agencies, a few went into forest firefighting, three were early ZeroG trainers for NASA. One of the firefighters was eventually used by Avco Lycoming for a jet engine test bed, YF 102 turbofan retracting into the bomb bay. When inspections found airframe life limits, it was donated to [NNAM]. A-2B BN 130418 is the last surviving Savage.

Tripping Over Engineering: Going Nuclear

AD-4B Skyraider
Service 1946 to 1973, foreign to 1985
Slowest nuclear bomber
3,180 built, all models

Skyraider - Douglas Sky Prefix & raider: one who attacks an enemy in the enemy's domain.

AD-1 with HVARs and Tiny Tims, barn doors open 1946 {NHHC}.

A war time design intended to replace other torpedo and dive bombers. Not entering service until after WWII, just about the last radial engine combat aircraft. In obsolescence when it entered service, Skyraiders shot down two Mig-17s on 20 June 1965 in Vietnam becoming the last piston engine aircraft to shoot down a jet fighter. It was also the first air-to-air gun kill of a fighter in Vietnam.

Operated by Navy, Air Force, Marines and eight other countries.   Operated in 28 distinct configurations:
- Airborne early warning with a radome and a four man crew.
- Anti-submarine warfare in hunter-killer pairs.
- Missile guidance aircraft for drone converted F6Fs.
- Four patient ambulance.
- Eight passenger transport.
- Carrier onboard delivery freighter (COD).

Tripping Over Engineering: Going Nuclear

- Delivering practically every form of ordnance, including a kitchen sink and a porcelain toilet.
- Attacking a dam with torpedoes
- If you punched out over Vietnam, an R-3350 was probably the next sound you wanted to hear after the chute popped.
- And THE SLOWEST US NUCLEAR BOMBER.

The introduction of the compact Mk-7 made a very large range of Navy and Air Force aircraft nuclear capable. The Skyraider was the first single engine nuclear armed aircraft, only the Navy deploying for nuclear weapons {Polmar 2009}. But, while large aircraft were enjoying inflight refueling, this was not yet common on fighter and strike aircraft. Fortunately, the efficient R-3350 with enough fuel and patience, the Skyraider could go a long way. It had only one defense, fly feet in the weeds and don't concede an opening as a target. Even though it was less of a worry at the time, it also pretty much kept you off the radar.

On nuclear training flights, there was a comment that departing the carrier deck at 85 ft was the highest point in the flight until setting up to land. These "butt busters" were flown single pilot, manual navigation, stick and rudder for 2,000 miles over 13.5 hours. But, for excitement there was always the weapon delivery. Two modifications were added to the AD-4B model for nuclear delivery. Structural reinforcement and a toss bombing computer. This permitted LABS delivery of the Mk-7. But the concept of LABS works best if you have the joy of high velocities with lots of inertia and at least reasonable thrust to carry you through vertical. At half the speed of jets it could only loft a bomb so high. One Skyraider pilot commented that on a LABS practice flight the Mk-7 dummy fired only 1,100 ft from the aircraft. {Michel 2003} {Miller 2001}

May 1953, an AD-4B set the weight record for single engine piston, 26,379 lbs take off weight with 10,500 lbs ordnance {Polmar 2009}. The Skyraider was applied as it always had, somewhat beyond its reasonable capabilities. The last of its kind, an aviation anachronism, performing in ways the original designers never anticipated. A stone age aircraft still useful in the iron age. There are a number of them on display and even flying the air show circuit. A beautiful one is at [NNAM]. {Rausa 1982}

Tripping Over Engineering: Going Nuclear

A3D Skywarrior
Service 1956-1991
1st jet carrier nuclear strategic bomber
282 built, all models

Skywarrior - Douglas Sky Prefix & warrior: a brave or experienced soldier

Two EA-3Bs over Gibraltar 1991 {NAC}.

The A3D was a replacement for the AJ Savage as the only two Navy carrier aircraft designed specifically as strategic bombers. Like the AJ, it was designed to carry a 10,000 lb nuclear bomb. It was the heaviest US operational carrier based aircraft at 84,000 lb. But the Douglas concept was the "lightweight" design, the original supercarrier bomber concept having been 100,000 lbs. In an engineering race for larger catapults and arresting gears, this limited the ships on which the A3D could be deployed. But like other aircraft, it was at a confluence of technology by the time it was in service. The Mk-7 permitted much smaller aircraft to deliver nukes and the strategic capability was far better served by ICBMs. Through its service the A3D was rated for 13 different types of nuclear ordnance {Silverstone 2013}. Its size and shape led to crews calling it "The Whale", but as a massive three crewed

Tripping Over Engineering: Going Nuclear

carrier aircraft without ejector seats its nomenclature was occasionally defined as "All 3 Dead".

The A3D had a full career with its size advantageous for hauling tonnage for tactical bombing and space for film photo reconnaissance. Some were converted to carrier VIP transports with a few operating as test mules for various aerial systems. The Navy even considered it as a heavy missile interceptor/airborne early warning for 90-200 mile stationing from the task force {Polmar 2009}. It was extensively used for F-14 radar development, even firing the Phoenix missile.

Operations like refuelling and ECM may not be glory jobs, but no carrier combat jock was ever unhappy to see a tanker or know ECM was overhead. In many ways it achieved the initial dream of having a heavy carrier bomber. It was used as a proof of excess demonstrator to show the Chinese its altitude capability, however marginal. While it normally had a 41,000 ft ceiling, with some flap and wedging into the coffin corner, it reached 52,000 ft {Miller 2001}. And as archaic as it was, it outserved it's high performance replacement by a decade. The A3D had a very long run of service, one of the longest serving carrier aircraft, but it was not a headline grabber like a lot of other aircraft.

While they are on the trolley ramp tour and not walk up accessible [NNAM] has two A3Ds. One was 24th of the initial 49 aircraft, originally a heavy bomber then a loft bombing trainer, ending its service as a weapons development aircraft. The other was initially a trainer then a VIP transport, to electronic counter measures combat missions in Desert Storm. It was assigned to Raytheon for weapons development and flew to Pensacola in 2011. It was the last flying A3D. [PASM] has one of the six YEA-3As that converted the bomb bay to a pressurized ECM space. Numerous others are on display.

Tripping Over Engineering: Going Nuclear

A-5 Vigilante

Service 1961-1979
156 produced
1st supersonic carrier nuclear bomber

Vigilante - a person who forces obedience to the law without legal authority to do so, or a member of a group that decides to force obedience to the law without official authority.

January 1968 RA-5C launch USS Ranger (CVA-61) {NHHC}.

Seems like a strange name for a nuclear bomber, one would hope that nuclear deployment would only be under legal authority. While this was a replacement of the A3D, it was no longer the concept of a strategic bomber. With smaller warheads, it was specifically considered for tactical operations. But where the A3D was early jets and late analog, the A-5 was new technology, much of which foreshadowed later aircraft. Large for its time, mach 2 capable, handling and looking more like a fighter than a bomber. But the engineering and operational costs of a supersonic strike aircraft was corrected in the move toward the A-6 and A-7.

Tripping Over Engineering: Going Nuclear

It was an early use of aluminum-lithium alloys and titanium. It had blown flaps, an early fly-by-wire system, heads up display, inertial navigation, multi-mode radar and television optics. Versatile Digital Analyzer "VERDAN" managed and controlled all systems, that is DIGITAL computing baby.

Technically impressive at so many levels, the most interesting was its design for handling nuclear stores. It provided fully internal space that could be operated at high speed without annoyances such as bay doors flapping in a supersonic breeze. A linear bomb bay was placed between the engines with a single Mk 28 sandwiched with fuel tanks and the tail cone. A cartridge would kick the entire package out the back, relieving itself over the target. The timing of ejection, chute retardation, air cushion deployment for laydown detonation and release into the highest turbulence zone of the aircraft precluded a precise drop of ordnance. Appreciating the consistent release of free air aerodynamics, A-5s were modified with pylons for more conventional strike operations. The Bedfellow "store train" was never used for functional weapons deployment. {Miller 2001} {Polmar 2009}

The bomb tunnel was considered for a few other purposes. A slide in aerial refueling system was considered, but with the A3D there was already a much more suitable aircraft. A sprint configured interceptor was considered in sliding a rocket into the bomb bay. One last concept was a high speed interceptor, stuffing a third engine into the bomb bay. Like the demise of a lot of equipment, it just lacked a proper convergence of technologies. One feature of Bedfellow was useful over its life. A linear bomb bay packed with fuel tanks. But as the tanks apparently retained the ejection capability, on more than one occasion the cat shot kicked off the tailcone and a tank. In one case resulting in loss of the aircraft.

On 13 December 1960 it set an altitude record with 1,000 kg payload that stood for 13 years, 91,450 ft. Air density of slightly over 2% of sea level. This is impressive even now, that ain't much air.

Like some of its predecessors, its speed and size were beneficial for tactical reconnaissance. Room for side looking

Tripping Over Engineering: Going Nuclear

radar, ECM and an extensive array of cameras. Apparently one camera configuration had a huge flash strobe for night photos. It was used a few times, discovering that photos were insignificantly improved and a big flashing light on a low altitude solo aircraft at night enticed a surprising amount of ground fire.

It's evolution to reconnaissance only probably lead to its eventual termination. Even with a few onboard a carrier, it still required support and maintenance of an extremely complex aircraft. Like some other cutting edge equipment, the costs related to the new technology led to an early demise where simpler aircraft could go on forever. According to one comment, it was an aircraft in search of a mission.

There are an array of aircraft on display, but [CAM] has the last Vigilante to carrier land on an overseas deployment. Two others I have seen are [PASM] and [NNAM]. And while some aircraft can't do it, you have to admit, just looks fast.

Tripping Over Engineering: Going Nuclear

CRUISE MISSILES

While everything in nuclear weapons started with bombers, all militaries want weapons that reduce crew exposure. The US had a range of missile development in WWII, but the Germans had led the missile and rocket charge with performance that could be considered the origins of strategic cruise and ballistic missiles.

The anticipated simplicity of cruise missile development in being unmanned aircraft did not match the guidance and flight control constraints of the time. Early systems required escort aircraft to remotely direct the missile. Better radars permitted accurate remote tracking with radio command guidance, but were dependent on significant RF bandwidth sensitive to detection and interference. Inertial guidance finally achieved the holy grail of full guidance autonomy with no RF bread crumbs.

The Air Force was considering the MB-47B using obsolescent B-47s for conversion to the world's largest cruise missile. While it was limited as a round trip bomber it had potential with one way range and heavy payload capability, achieving the HALFMOON plan without having to expend all of the crews.

The development of defenses against ICBMs lead to a return of cruise missiles. But improvements in all technologies lead to the modern cruise missile being much less observable and more capable of low altitude flight autonomy. The modern military drone is simply a recoverable cruise missile with radio command guidance. With a bit of knowledge and a bit of money, anyone can build a cruise missile. Modern aircraft have a profusion of the capability already built in.

TDR-1

Operational 1944
195 aircraft built of 2,000 ordered
Maybe the 1st carrier launched guided missile [NNAM] has the only survivor

TDR-1 preparation for Rabaul attack Sep or Oct 1944 {NHHC}.

There were wartime expedients to the cruise missile. Aphrodite and Anvil programs used poor condition B-17s and PB4Ys equipped with autopilots and radio control gear. The aircraft were stripped of crew, defenses and non essential equipment so the only concern was range, tonnage and control. An aircraft stripped for a one way flight went from a 2,000 lb bomb load to a 30,000 lb explosive load without concern for bomb casings and release gear, just pack it full. A crew on board would perform take off, the escort aircraft taking over on remote and the crew bailing out over a predesignated area. At the target zone the remote pilot

Tripping Over Engineering: Going Nuclear

would simply fly the aircraft into the target with a relatively new vision system called "television". Fourteen missions were flown, with the deaths of several take off crews (the most visible being Joseph Kennedy Jr.) and no successful target destruction. This lead to a definitive termination of operation in 1945.

The Navy wanted a precision weapon that could be used from carriers at remote islands in the Pacific with less risk than the crewed takeoff of Aphrodite. The first TDN design was complicated and expensive for a disposable aircraft, leading to the TDR-1. The manufacturing resources were outside of typical military industrial production. Schwinn Bicycle welded a lightweight tubular steel frame, almost an homage to the Wright brothers. Wurlitzer Music Company laminated and molded complex wood shapes, replacing pressed aluminum. A pair of 220 HP Lycomings, minimal strategic materials use, 2,000 lb bomb capacity, optional ferry cockpit and the latest in radar altimetry, RCA television and radio control. A TBM control aircraft monitored the TV which also viewed critical flight instruments. An autopilot system kept it straight and level. A telemetry system also provided for ten preset commands such as flying at fifty or one hundred feet altitude. This was achieved through the simple radio telegraphing of a rotary dial phone system. Nothing like phoning your plane with commands. In the Pacific, 50 aircraft were launched with 31 hitting specified targets.

It was a true wartime expedient, minimal expense to accomplish the mission. In Korea, six F6Fs equipped with improved television systems were controlled by Skyraiders, three hitting critical precision targets. The Walleye in the 1960s finally achieved fire and forget capability with self guiding TV targeting. The TDR-1 could be considered both a drone and a guided missile. A significant limit was the line of sight for the radio of the time. With a high altitude relay aircraft or satellite, it wouldn't be all that different from the Predator. This is the military drone legacy, all it needed was better controls and communications.

Tripping Over Engineering: Going Nuclear

JB-2 Loon
Testing 1944-1949
1,391 missiles built
1st submarine launched guided missile
Loon - an aquatic bird, now on the Canadian dollar coin

First Loon launch from USS Cusk 12 February 1947 {NHHC}.

On 6 June 1944 the massive D-Day landings assaulted German occupied France. 13 June 1944 The Germans started firing accumulated V-1 missiles on Britain and Belgium. Of 8,654 fired only 2,419 hit Britain and 2,448 on Belgium. With the invasion, the US had enough in possession within a month of first launch to copy the missile as the JB-2 Loon. Both the Army and Navy intended to use them for a planned assault on the Japanese home islands at a firing rate of 500 per day. While the atomic bombs ended the war before deployment, it remained a significant interest {Polmar 2004}.

The Germans had launched from fixed base operations, but the US knew that mobility would be required in the Pacific. The Army experimented with launching from B-17s and B-29s to provide a standoff weapon from bombers. The Navy tried PB4Y aerial launches and LST/CVE ship launch systems. One fired from the USS Cusk (SS-348) on 12 February 1947 was the first guided missile launched from a submarine. The accuracy of the original V-1 guidance was based on altitude and heading autopilot with a nose propeller that simply counted rotations to a preset distance at which it pitched over onto target. This required best forecast adjustments for any wind conditions and it accumulated heading errors over the range. But, it was a fairly economical missile able to deliver an 1,830 lb warhead 150 miles into an area target. The Loon was improved with radio-command guidance requiring course correction from the control vessel {Polmar 2004}. The

Tripping Over Engineering: Going Nuclear

German program had considered radio guidance for precision, but the large city targets allowed the minimally self guided missiles to hit within seven miles of target in the later firings.

With the delay of the Regulus, the Loon had been considered as a missile for a 3 KtE Mk-8 warhead. The range was considered inadequate and the accuracy unsatisfactory for only a 3 KtE warhead. While the Loon ended life only as a development weapon, it lead to a plan for Navy and Air Force cruise missiles {Polmar 2009}.

Several on display at [AFAM], [NMUSAF], [NASMuh], [WSMR] & [EASM]. A guidance unit is shown at [NMMSH].

Tripping Over Engineering: Going Nuclear

MGM-1 Matador
Service 1952 - 1962
~1200 missiles built
1st US nuclear cruise missile
Matador - A bullfighter whose task is to kill the bull.

The Loon laid a groundwork for cruise missiles, but had significant limits from its original intent. Fact of life, the original V-1 was a quick cheap method of throwing explosives. Its only requirement was hitting a city with an abundance of missiles, tolerating a significant level of attrition. Once you go to the expense of mounting a nuclear warhead, you start mandating more specific requirements on the targeting capability.

The Matador was the first operational nuclear cruise missile of the US. At the time of introduction, it was designated B-61 even though it previously had a missile designation. Yep, it was called a pilotless bomber and the initial units were pilotless bomber squadrons. It went into service with the W5 20 KtE warhead and a 2,000 lb conventional warhead. It was eventually designated as a tactical missile which questions its operation within the Air Force. With mobile launchers hauling a 39 ft long, 28 ft span, 12,000 lb aircraft and a 620 mile range, it necessitated deployment near the theatre of operations in a more Army like fashion. The J33 jet was a great improvement in performance and efficiency over the Loon pulsejet. Besides better power and control range, it was an established fleet engine. A big Aerojet solid rocket booster kicked it off the launcher to flying speed. Staying subsonic provided a great aerodynamic simplicity, but with the terminal dive, the transonics caused some twitchy control issues.

Guidance at the time integrated analog computers with ground radar stations to gave radio command guidance enroute. Banging away with radar and a continuous stream radio communications. Good conditions had about a 250 mile range with control handoffs possible, but inconsistent in performance. Trainers and fighters were sometimes equipped with the missile control system to fly over the missile site onto a typical launch path so the crews could train

Tripping Over Engineering: Going Nuclear

on live guidance. Considering the EMP and RF noise of a nuclear detonation, it questions the capability of a nuclear counterstrike. Later versions were still tethered to a line of site microwave signal, claiming extended navigation range, but requiring a significant transmitter network to achieve it. Shanicle (Short Range Navigation Vehicle) guidance was a harbinger of the future. LORAN had used lower frequency to achieve over the horizon capability with constrained precision, but it operated very simply on a time range system. Multiple transmitters are slaved to a master with the arrival of signals from different stations providing position based on the time distribution. The Shanicle system used similar time based system, but with microwave transmitters and a much higher accuracy. GPS extensively performs the same function from geostationary satellites.

Matador almost looks like a mini fighter with the swept wings and T-tail. The T-tail had to make the control dynamics much simpler than the Snark or Regulus. With the NACA duct flush with the belly and the finger roll spoilers, it looks aerodynamically elegant.

Multiple typically on pylons [NASMuh], [NMNSH], [NMUSAF] & [MOA].

SSM-N-8 Regulus I
514 missiles built
Service 1955-1964
1st nuclear Navy missile
1st operational submarine launched missile

Regulus - Brightest star in constellation Leo, Formerly a petty king

Regulus I in launch position, 26 August 1954 USS Tunny (SSG-282), The first US nuclear missile submarine {NHHC}.

Regulus I subsonic cruise missile was intended for a 4,000 lb conventional warhead, but by 1949 the plan was focused on nuclear with a 3,000 lb warhead and 575 mile range. The TROUNCE guidance system required remote guidance, but it was demonstrated over a 313 mile path handing off between 3 guidance vessels to strike within 150 yards of target. It used the vessel air search radar to provide integrated control with vector data. It was initially tested onboard cruisers, but with folding surfaces to fit in tiny hangars built on submarines. The J33 jet engine provided better behaved start and run capabilities than the Loon pulsejet or the original preferred ramjet and was already used throughout the early jet fleet. Since it was boosted to adequate flying speed and consumables were configured for limited shift of center of gravity, the swept wing allowed pairing the ailerons to operate as elevons. No horizontal stabilizer was needed and only a vestigial vertical stabilizer/rudder. These aerodynamic efficiencies would also arise on the Snark.

It transitioned from the W5 50 KtE warhead to the W27 with 2 MtE. The Navy program office at one point considered it an unmanned aircraft rather than a missile. There was a

Tripping Over Engineering: Going Nuclear

consideration of the USS Nautilus being Regulus equipped, but Rickover insisted on it being a test of nuclear propulsion, without a lot of other experimental systems that could potentially deem the entire nuclear submarine program a failure.

The initial Regulus submarines Tunny and Barbero were wartime diesel electric fleet boat conversions. Both had two missile hangers and launch rails behind the sail. Tunny would launch the first Regulus on 15 July 1953. The Halibut was initially intended as a diesel electric specifically designed for missile launching, but was changed to nuclear in construction. The Growler was converted to missiles during construction. This made Halibut the first nuclear powered nuclear missile submarine.

The introduction of the Polaris SLBM rapidly illustrated the limitations of the Regulus. Since these missiles required a surface launch, some surface ships were also tested as missile carriers. Four Baltimore class heavy cruisers were configured and deployed with Regulus. With new jet carrier aircraft performance taxing the ability of the remaining wartime carriers, a new option was tested. Several older carriers were tested and a few deployed with Regulus. Compared to a rail and rocket launch, carriers provided an elegant solution. A reinforced trolley made for convenient movement of the missile around the carrier. When ready to launch, ride up to the flight deck, hook it up to the cat, shoot it off the bow, trolley falls away and Regulus saunters on its happy way. Witnesses specifically commented on how well behaved it was on cat launches {Polmar 2004}.

The SSM-N-9 Regulus II was developed along the lines of all aircraft and cruise missiles at the time; larger, faster, higher, farther. Mach 2 at 59,000 ft for 1,100 miles, also 23,000 lbs and 58 ft long, doubling the performance of Regulus I with full inertial navigation allowing autonomous and independent operation. To a large extent it achieved the proposed performance, but like the other cruise missiles, the cost of this performance could not compete with the ability of the new ballistic missiles. Canceled in 1958 with 54 produced. Depleted as target missiles, especially for BOMARC supersonic testing, only three are retained on display.

Tripping Over Engineering: Going Nuclear

The Navy and Army were in contention on the Regulus I versus the Matador, but the Regulus had specific requirements for Naval purposes. Some were equipped with landing gear for development testing which was used later on the target drones. The Regulus systems were instrumental in the growth and development of later cruise missiles. The Regulus I after dropping its launch gear looks like the ultimate aerial vehicle; intake at the front, nozzle at the back, minimal aerosurfaces, a jet engine with wings.

USS Growler (SSG-577) was in service 1958-1964. The [ISASM] has the Growler with a Regulus I in launch position. [NASMuh] has one sitting on launch rails. [CaAM] has one sitting on the carrier trolley, amazingly simple for something shooting off a giant slingshot.

Tripping Over Engineering: Going Nuclear

MGM-13 Mace
Over 1,000 produced
Service 1959 - 1970s
Mace - A club type blunt weapon with a heavy head.

Mace B on TeraCruzer trailer at [AFAM] {LB}.

The Mace was a very direct descendant of the Matador at 18,750 lbs, 45 ft long and 23 ft span, giving more capacity for upgrades. The wings now folded instead of unbolted for quicker mobile launch preparation with a W28 thermonuclear warhead up to 1.45 MtE. It had a low altitude range almost equaling the Matador's at altitude while providing double the Matador high altitude range to 1,500 miles.

Tripping Over Engineering: Going Nuclear

The range increase provided the option of fixed installations in European based coffin launchers while still maintaining the flexibility of mobile launchers. With the distinctly longer heavier Mace, the coffin launchers also reduced the hauling of these monsters. Where Matador had more conventional tractor trailer configured transporter launchers, the Mace received the TeraCruzer system to provide a low ground pressure system for something hinting at battlefield mobility. Since freeway cruising isn't the forte of a 45 ft missile, the reduced roadworthiness would not have been missed.

Both Mace and Matador had an interesting approach to the launcher that compacted the linkage for shifting transport and launch positions. Rather than a full sliding rail that potentially restricts motion at the end of launch, these had a three point mount that allowed the forward section to drop clear upon initial movement. Lacking runout, it also required more critical attention to booster thrust minimums and center of gravity alignment.

With the range increase, particularly at low altitudes, the Matador radio command guidance and microwave navigation systems were distinctly limited. Bombers had been operating with manual radar navigation and bombing systems since the B-29, but this required automation for missile application. Mace A used the ATRAN system which matched the missile radar to a radar profile recorded on 35mm film through two AM signals. The magical concept of photographic film as a memory device for an analog computer. Initially, the only way to make the film map was by flying an aircraft over the planned path with a fixed angle radar, attesting to the extensive use of spy aircraft at the time. The Mace B received a proper inertial guidance system that provided non RF autonomy.

The Mace had the same elegance as the Matador, just longer and leaner. Multiple on display, typically on pylons with some at [WSMR], [NMUSAF], [NMNSH], [MOA] & [AFAM] has one on the all terrain launch trailer.

Tripping Over Engineering: Going Nuclear

AGM-28 Hound Dog

722 produced
Service 1960 - 1975
1st guided nuclear standoff missile

Hound Dog - Hit song performed by Willie Mae "Big Mama" Thornton and to a lesser extent by a minor entertainer named Elvis Aaron Presley.

Hound Dog mounted on B-52 {NMUSAF}.

In the race for supersonic intercontinental cruise missile was the land based SM-64 Navaho. When the intent is Mach 3 at 77,000 ft, a ramjet gets to be a pretty good idea. It also means that you have to get 65,000 lbs up towards Mach 1 when the ramjet really starts cooking. Fortunately, it means that you don't worry about minimum control or lift airspeeds. By the time the engine is kicking in, you have all the lift you want. While the intercontinental cruise missiles were found to be problematic, the kerosene/LOx booster on the Navaho significantly influenced the design of the ballistic missiles that put it out to pasture. The only surviving Navaho is currently undergoing restoration at the Air Force Space & Missile Museum, Patrick AFB FL.

But, lessons of the Navaho did educate other development. A large bomber investment drove interest in standoff weapons

Tripping Over Engineering: Going Nuclear

made more significant with the increased Soviet anti-aircraft capability. With large anti-aircraft missiles in fairly soft fixed base operations, a simple area guided missile with enough yield could eliminate the threat to the bomber. Carrying a W28 warhead at 1.45 MtE, the Hound Dog was a significant strategic weapon in its own right. Intended as an interim weapon, it served temporarily for fifteen years. While air launch took away some of the issues of Navaho, a B-52 now had to haul two 10,000 lb pylon mounted aircraft. The use of J52 engines with extensive fleet history eliminated a lot of unknowns. Capable of operating over Mach 2 at altitude, it had a number of attack configurations. At 56,000 ft with a max range of 785 miles exposed the highest radar visibility, but that also gave the ability to decoy if it was the greater requirement. It could operate clear of obstruction at low altitude or terrain following on radar down to one hundred feet. Dogleg targeting allowed flight to a waypoint as a distraction before turning toward target. As long as the B-52 was over 5,000 ft they could launch.

The Hound Dog internal navigation was inertial, but an effort to provide the best location datum at release was through celestial navigation integrated in the pylon. This was placed onboard the missile in later models. Experiments were made with terrain contour matching navigation and even with a passive radar detection system for use as a radar killer. Most weapons are dead weight until fired, but Hound Dog provided a shared resource with interrogation of the celestial navigation by the aircraft to correlate positional errors. Having a turbojet with a proven reliability, another peculiarity was integrated to assist takeoff. Missile engine operation and fuel connections were provided so the bomber had 15,000 lbs of thrust available from the missiles, making a ten engined B-52. But, with a low mounted engine to vacuum the runway and sitting in the splash and trash zone of the nose gear, damage to the missile was always a risk. The engine was also configured for maximum thrust over a relatively short life, so it was always preferred not to run the missile for assistance {Gibson 1996}.

An aircraft with an engine larger than the fuselage and hardly any visible means of support looks irrepressible. Never used in combat, never retasked for other missions, just retired and scrapped. Some are on display at [CAM], [AFAM], [NMUSAF], [PASM], [MOA] & [SDASM]. [NMNSH] has one under a B-52 on the loading trailer in the typical flight mounted position.

The Hound Dog flew with a little brother that didn't quite share the looks. It was not unusual to carry a couple of ADM-20 Quails tucked into the bomb bay of a B-52. These little stubby jets would be dropped as decoys with a wing span of only 5-1/4 ft with a 13 ft long boxy fuselage. Flying along at B-52 altitudes and speeds for 445 miles. Radar reflectors and other systems enhancing visibility of the little bird. On radar appearing for all practical purposes like a B-52. The Quail only lived to be shot so all of the other weapons could reach the target. Of 585 built, a fair number are still around. Some are at [PASM], [SDASM], [NMUSAF] & [MOA].

Tripping Over Engineering: Going Nuclear

SM-62 Snark
30 missiles active less than four months in 1961
1st intercontinental cruise missile

Snark - "For the snark was a boojum, you see", from "The Hunting of the Snark" by Lewis Carroll, a fascination of Jack Northrup.

Cape Canaveral test launch 1956 {NMUSAF}.

The first intercontinental missile solution was not originally a question of ballistics {Dobbs 2008}. There was plenty of experience in long range aircraft, so extensive technology was available. Wrap a fuselage around an existing J57 jet and 3.8 MtE can reach out and touch someone at 6,300 miles. For a variety of reasons, the ONLY US intercontinental cruise missile.

With the long range and the potential airborne alert capability, it integrated one of the few missile retrieval features. Return to a designated landing area and belly down on a runway, not a quick turn around system. A subsonic aircraft rocking along at 50,000 ft with no countermeasures eventually becomes a pretty easy target. The terminal warhead delivery did provide some defensive capability in target engagement. Instead of flying direct to detonation point, the warhead would separate with a fuselage pitch up for clearance and the warhead aero package free falling to target. For a sacrificial airframe, this appears to be an unnecessary complication, but it did have a few benefits. It eliminated the pesky transonic

Tripping Over Engineering: Going Nuclear

conditions of a subsonic aircraft in terminal dive and it would have been harder to hit the warhead package with the flight unit flailing like a windmill on radar.

The long lean fuselage and the minimal tail surfaces give this 67 ft long, 42 ft span 60,000 lb monster a strangely svelte beauty. The intake plenum gave it a bit more curve in the backside, placing more lateral surface aft requiring only a sliver of a vertical stabilizer. In a four month period in 1961 thirty service missiles were commissioned and canceled at Presque Isle ME, its only missile base {Polmar 1975}.

Steering by the stars, celestial navigation gave it stealthier autonomous navigation, but a test launched from Canaveral with targeting off Puerto Rico was eventually found in Brazil years later. Testing was so problematic that some crews referred to Canaveral as "Snark infested water".

The SSM-A-5 Boojum was the intended heir to Snark. Bump the speed up to mach 2, add another 20,000 ft to altitude with the logistical convenience of 85 ft long, 51 ft span and a feathery gross weight of 112,000 lbs. Designed for launch by rocket sled with consideration as an air launched cruise missile from the B-36. As if there weren't enough problems with the Snark design.

Five Snarks remain at [NMNSH], [SACSM], Air Force Space & Missile Museum, Hill Aerospace Museum & [NMUSAF] has one sitting on the launcher.

Tripping Over Engineering: Going Nuclear

BALLISTIC MISSILES

The Germans had put major work into the V-2 during the war. LOx (liquid oxygen) and alcohol/water tossing a 2,200 lb warhead 128 miles. Rudimentary inertial guidance was adequate for an area weapon, even a concept for mobile submarine launching.

The scientists, engineers and equipment that developed the V-2 were the core of missile and rocket development for the US and USSR after the war. On 6 September 1947, USS Midway became the only moving V-2 launcher. Of all carrier launches, standing up a rocket with the annoyance of LOx loading on the flight deck, blast off would be spectacular. The Navy was even considering V-2 missile conversions for a battleship and a cruiser {Polmar 2004}. Multiple V-2s are on display but [WSMR] has a cutaway.

For nuclear delivery, ballistics became the holy grail. Rather than sending bombers or cruise missiles faster and higher, just boost the warhead into space and let it free fall with high velocities, short flight time and a barely observable warhead package greatly short circuiting the anti-aircraft defenses.

Ballistics eventually provided ranges from a few miles to halfway around the world. Once the guidance systems had reasonable accuracy and reliability, the launch vehicle was an issue of scalability. Get enough impulse and you can throw as heavy as you want as far as you want. Five minutes of burn time and half hour of flight time. Solid rockets achieved the previously elusive goal of storable ready to fire missiles.

Better warheads meant more delivered yield or smaller missiles. Once terminal guidance was achieved, the development of the MIRV lead to smaller, better directed weapons. Even with ABM capabilities, the strategic weapons core is still ICBM. Throw enough warheads and they can't stop them all.

Tripping Over Engineering: Going Nuclear

ARMY BALLISTIC MISSILES

When it comes to ballistic missiles, the discussion is boomers in the Navy and silos in the Air Force. Little is thought of extensive parallel development done by the Army. The first American in space was not on an Air Force rocket, it was an Army rocket. Army work on solid rockets had direct relationships with the Navy SLBM program. The Army kept nuclear ballistics stationed in Europe through 1991 with up to an 1,100 mile range.

MGR-1 Honest John Missile
7,000+ missiles
Service 1953 to 1991
1st nuclear capable surface-to-surface rocket

Honest John - Supposedly the name given by Colonel Toftoy when the accuracy of the rocket was questioned.

Honest John double launch with different stages of spinner ignition {RA}.

Honest John is credited as the first nuclear capable surface-to-surface unguided rocket. Originally with the W7 warhead, later changed to a 2, 20 or 40 KtE W31 boosted fission shared with the Nike-Hercules. This was truly a tactical missile with a 15 mile and later a 30 mile range. To accurately toss a dumb 2.5 ton device required a rail system with a lot of thrust (92,500 to 150,000 lb thrust) and a very short burn. While the fins provided stability and some spin for accuracy, the flight time was so short that spin needed initiation at launch. The head of the rocket has eight tangential nozzles that shoot flame laterally once clear of the rails. An elegant solution that momentarily looks like the motor is disintegrating. The

Tripping Over Engineering: Going Nuclear

inside/outside rail system provides a shorter slide on loading and a simultaneous fore and aft rail release. The truck launcher is powered and integrated, but the towed launcher is pure Meccano with a lot of shafts and screws terminating at azimuth and elevation hand wheels under the nozzle of the rocket. Sitting under the nozzle while aiming would significantly reinforce the following of firing safety controls by the artilleryman.

Once on the mobile or towed launcher it looks very simple, but an array of equipment was required to assemble and mount the three shipping components. It is preferred to ship explosive nuclear parts separate from hot burning parts. While there was also an optional 1,500 lb conventional warhead, another was far more perilous. A shared warhead with the Corporal contained 356 Sarin nerve gas submunitions, each with its own containment and bursting charge. Considering that Sarin is colorless and tasteless, I would rather handle the nuke.

Honest John had an extremely long service life, well liked for its simplicity. It did what was needed and left little need for upgrades. It is probably one of the more recognized rocket shapes of its time. The Army deployed it as artillery and the point and shoot ballistic profile provided a very artillery like operation and accuracy considering the range. If you are not going for multishot delivery efficiency, a rocket gets rid of all of that irksome gun weight. At least gun artillery crews didn't have to wrap rockets with electric blankets to ensure maximum temperature uniformity for consistent firing.

Honest John did have progeny which warrants some technical discussion. It was a light weight air transportable complement called, wait for it, Little John (MGR-3). 500 were built for service from 1961 through 1970. It was a dumb free flight artillery rocket with an airmobile weight of 780 lbs, 11 mile range and 1 to 10 KtE yield. It would not be excessively notable except for one accuracy enhancement over Honest John. Rather than post launch spin rockets, it was considered necessary for the rocket to be spinning when it left the rails. This became the SOSR, spin-on-straight rail system. The missile would be prepared, armed and laid. A crank wound up a coil spring 8.5 turns. With the rear launch shoe on a

Tripping Over Engineering: Going Nuclear

swiveling fin assembly, the spring drive connected to the main body of the rocket. In order to fire, the firing lanyard released the spring drive spinning the rocket up to 210 rpm, at which point a centrifugal internal switch would fire the main rocket {Keller 1960}. The Army was the only service using a wind up missile.

There are a range of Honest Johns on display including [USSRC], [WSMR] & [CaAM]. [NMNSH] has both truck and towed launcher units. Little Johns are around, but finding one intact with the spring spinner is more problematic.

Tripping Over Engineering: Going Nuclear

MGM-5 Corporal Missile
1,100 missiles built all models
Service 1955 to 1964
1st US guided ballistic nuclear missile
Corporal - Army rank just above private

Corporal on transporter {RA}.

Corporal was an interim project from the WAC Corporal. This had a pressurized tank feed of red fuming nitric acid & aniline with 20% furfuryl alcohol motor and Tiny Tim boosters. The Tiny Tim was a WWII Navy aircraft anti-shipping rocket. Per its name, it was over ten feet long, casing just under 12 inch diameter and weighing 1,255 lbs. Who said the Navy doesn't have a sense of humor. The Tiny Tim boosters gave a 0.6 second kick of 37 Gs. Without any active control, the only way to provide stability was to get to a speed where the fins were aerodynamically adequate.

Corporal was the first US nuclear ballistic guided missile with

Tripping Over Engineering: Going Nuclear

an 86 mile range and a W-7 warhead. Purchases by the UK made it the first foreign owned US guided missile. Firing red fuming nitric acid and hydrazine can be seen as a step toward the Titan II propellant combination. These had a range of chemical, temperature and corrosion sensitivities, but unlike Titan required field mobility with fueling after assembly and launch stacking. At 45 ft height and launch weight of 11,250 lbs, it required seven hours for launch set up. One missile would deploy with a surreptitious entourage of 15 vehicles and 250 men {Polmar 2009}.

Guidance was inertial during boost to maintain orientation and attitude. Then a radar/transponder system guided through radio command requiring line of sight and a lot of RF noise to direct the missile. All control is through basic vane and rudder. While the hot area vanes are carbon for temperature tolerance, the hinge tabs and cable pulls don't look all that different from the rudder controls on a Cessna 150.

The original Corporal I improved through three II models. There was a plan for the III, but the Sergeant was at adequate progress to replace Corporal. Solid fuel greatly simplified handling and storage requirements {Bragg 1961}. Several are around, including [WSMR] & [USSRC].

PGM-11 Redstone Missile

128 Built
Deployed 1958 - 1964
1st thermonuclear capable ballistic missile
1st missile delivered nuclear test

Redstone - Named for Redstone Arsenal, which is named for the red clay soil.

Redstone 1st tactical firing with part of its extensive retinue, 6 July 1961 {RA}.

Redstone was the first missile capable of launching a thermonuclear warhead. It was considered a tactical missile, but at 3.5 MtE it had three times the yield of the largest current US strategic warhead. With a max range of 200 miles, it did require a level of "mobility". An Army field artillery school document from January 1959 shows a train of 12 trucks with 11 trailers to dispatch a Redstone. Part of this train was carrying anhydrous hydrogen peroxide, liquid oxygen and eventually hydrazine. Included was a 25 ton mobile crane which was later replaced by a combined launch platform with A-frame.

Tripping Over Engineering: Going Nuclear

The first and second missile delivered nuclear tests were by Redstones on Hardtack Teak and Orange. Teak was 1 August 1958 with 3.8 MtE detonated at 252,000 ft over Johnston Island. Orange was 12 August 1958 and detonated at 141,000 ft. These were intended to ascertain the use of high altitude nuclear bursts for killing inbound ballistic missiles, particularly with EMP and X-ray burst. The Teak shot was visible for a half hour 800 miles away in Hawaii. The intended effect on RF signals was notable in that Johnston Island was unable to contact command in the states on HF for eight hours. While Orange did not have the same level of interruption, after Teak, they notified Hawaii of a nuclear test window. There was the reporting of an insignificant issue in Hawaii of airliners losing contact with air controllers. But in those days, aircraft were still mechanical and you had flight crews that a few years before had been using sextants.

The fuel was a carryover from the V-2; 75% ethanol, 25% water, smidge of purple die and a soupçon of methyl alcohol to stop those pesky corporals from tippling out of the fuel tank. For all of the issues in boosting specific impulse, there was probably significant discussion that 75% alcohol and 75% range is way better than 100% alcohol and burning through the motor on the pad. The Redstone was an early change to the plate injector motor giving better combustion and heat management. In the era before bootstrapping, catalytically reacted hydrogen peroxide provided hot start of the pumps.

In an effort to increase specific impulse, Rocketdyne's Mary Sherman Morgan was the technical lead developing Hydyne. This was an early development in hydrazine fuels with 60% unsymmetrical dimethylhydrazine and 40% diethylenetriamine. This would be a significant influence on the Aerozine 50 used on Titan IIs. Besides the cryogenic and oxidative effects of LOx, now there was hydrazine toxicity to worry about.

The Redstone also had a properly self contained inertial guidance unit. Given the short flight time, there was far less time for that dearth of inertials, accumulated drift. The gyro table had a limited the range of motion to index the accelerometers. Therefore, the preference was to limit the range of motion and specifically position the rocket. Once the

Tripping Over Engineering: Going Nuclear

launch platform was bedded and leveled, the rocket was positioned. The launch platform fully swivelled so that theodolites could position the azimuth with Fin I lined up on the target within 30 seconds of rotation. The heavy computing was in the control truck which uploaded a ballistic profile to the missile. This is a far cry from punching a lat-long and letting it fly. The boost control was azimuth and pitch with engine cutoff for range. Air puffers provided high altitude attitude control with aero surfaces for terminal control.

Redstone provided a significant amount of education on rocketry and even though its field deployment was shorter than its development, direct derivatives gave the US its first steps into space, launching the first two Americans and the first US satellite. Replaced in service by the solid fueled Pershing to the joy of logisticians. With its retirement, a range of rocket testing and satellite launches were provided by Redstones. One military derivative considered was the Redstone freighter with the ability to fire a supply capsule isolated ground troops. Why not just take this to the conclusion of rocket deployed troops with a full circle back to Starship Troopers {Heinlein 1959}.

There are multiple Redstones on display but [USSRC] gets you into Redstone Arsenal. [USSRC] also has a range of Redstone components on display. [WSMR] has one on its launcher with the transport trailer and several support trailers.

Tripping Over Engineering: Going Nuclear

AIR FORCE BALLISTIC MISSILES

In 1953 Colonel Bernard Schriever declared to his former commanding officer Curtis LeMay, that the mission of the B-52 would be much better served with a modified B-47. In 1954 he was put in charge of all Air Force ballistic missile programs {Schwiebert 1965}. He became the godfather of the ICBM in the same way as Groves for the atomic bomb and LeMay for strategic bombing.

Schriever promoted extensively to get space program funding for the missile program on the thought that the missiles would be the core of any space program. He finally received funding which was not allowed to be used for any component or system development. After a speech in February 1957 addressing the missile program as the foundation of progress into space, the next day received orders never to use the word "space" in any speeches {Schwiebert 1965}.

In the early ICBM days, there were repetitive launch failures. This would be expected with massive quantity of explosive fuels expended entirely in the first five minutes of flight in an entirely new device. These failures reinforced continued support for bomber and cruise missile programs. Within ten years, engineers who had worked aircraft and ICBM programs considered the ICBM to be an easier engineering exercise than a bomber {Schwiebert 1965}. ICBMs required high reliability, but only had to work a half hour. Where a bomber or cruise missile was a relatively slow and low target for a long time, early ICBMs were pretty much untouchable after launch.

Air Force ICBM timeline

The timeline demonstrates the extensive parallel Air Force development in the ICBM program.   Also, the points of convergence on the more effective systems.
Propellant types:
CL - Cryogenic Liquid
SL - Storable Liquid
SF - Solid Fuel

1954 May CL - Atlas declared highest priority AF project.
1955 May CL - Start Titan development.
1959 Feb SF - Start Minuteman development.
1959 Oct SL - Start Titan II development.
1959 Dec CL - 1st operational Atlas base.
1962 May CL - 1st operational Titan I base.
1962 Oct SF - 1st operational Minuteman I base.
1962 Oct/Nov - Cuban Missile Crisis
1963 Mar SL - 1st operational Titan II base.
1965 June CL - Last alert Atlas.
1965 June SL - Titan II replaces Titan I.
1966 Apr SF - 1st operational Minuteman II base.
1970 Dec SF - 1st operational Minuteman III base.
1972 Apr SF - Peacekeeper development starts.
1986 Dec SF - 1st operational Peacekeeper base.
1987 Aug SL - Last alert Titan II.
1993 Dec SF - START I treaty destruction of MM II silos.
2005 Sep SF - Last alert Peacekeeper.
SF Minuteman III only remaining land based ICBM.

Tripping Over Engineering: Going Nuclear

SM-65 Atlas Missile
350 produced all models
Only D/E/F were ICBMs
ICBM service 1959 to 1965
First US ICBM

Atlas - (Greek) Atlas revolted against the gods and was sentenced to bear the heavens for all eternity.

Atlas Launch {NMUSAF}.

Tripping Over Engineering: Going Nuclear

The Atlas was originally intended to be much larger, but the forecast weight reduction of warheads slimmed down the design. Being very early in the business, there was a particular concern on restarting a rocket in space, so rather than staging they decided to burn them all. Two booster, one sustainer and two verniers all fired at launch. After initial boost, the two boosters were dropped with the sustainer burning until cut off, commonly known as stage and a half. The verniers would provide final control until warhead separation. Besides not worrying about space ignition of the LOx/Kerosene motor, it also provided a +G condition all the way to cut off. This ensured that propellants were always loaded to the pickups. In order to minimize the tank mass, the issue was not shell strength but rigidity. The thin wall sections could easily get into buckling mode without an extensive amount of reinforcement. Taking advantage of basic tensile strength, the tanks were pressurized to five psi under all conditions {Lonnquest 1996}, even when empty.

Depending on configuration, there was a 6,400 to 9,000 mile range. A, B & C were all test models with various levels of guidance and control. D was a soft coffin launcher with radio inertial guidance and a 1.4 MtE warhead. E & F had full inertial guidance and a 4 MtE warhead in hardened installations with E in coffins and F in silos for higher overpressure. The fueling process required fifteen minutes. In the case of D & E, a very long fifteen minutes, not including lift time, standing tall and proud above a coffin launcher with the sky full of inbound bombers and missiles. Totalizer slowly ticking over as "We'll Meet Again" plays in the distance.

The F model silo configuration allowed fueling in the covered silo with a two minute ride to the surface for launch. The trepidation of loading LOx into a warm tank in an enclosed space, fogging oxygen vents like the smoking nostrils of a dragon in its cave. The need for high levels of ventilation does not marry well with the needs of a hardened installation. The only operational explosions of Atlas missiles were four F sites during fueling {Gibson 1996}.

An "improvement" was the Quick Fire program providing the ability to launch an entire Atlas force within five minutes of a

Tripping Over Engineering: Going Nuclear

launch order. The crews simply had to keep the missile fully fueled in the silo. Vaporing O2 from a minimally insulated cryo tank frosting over with any humidity. For short periods of time, a launch ready ICBM, but in no way a storable launch ready.

The first ICBM had a very short service life with improved missiles, but it already had close association with space systems. 18 December 1958, a four ton communications satellite was launched via Atlas. A christmas message from President Eisenhower became the first satellite relayed broadcast. John Glenn was established as the first American in orbit launching 20 February 1962 on an Atlas. Atlas would launch three more orbital astronauts. Mariner II got a ride as the first spacecraft to successfully achieve contact with another planet from an Atlas-Agena-B on 27 August 1962 {Schwiebert 1965}. Atlas launched the Ranger moon surveys and the Gemini intercept targets in preparation for Apollo. And as a satellite launch system, Atlas derivative launches are scheduled out to 2020. Maybe Atlas truly was destined to bear the heavens for eternity.

Gleaming stainless steel 83 ft tall, 267,000 lbs ready to rock. A Mercury Atlas on a transporter at [USSRC] is much more accessible than the vertical mounts. There is one standing at [SACSM]. A gyro/accelerometer package is at [NMMSH]. Most of the launch sites were decommissioned and sold to the public with various conditions and accessibility. In some cases you may be able to contact a landowner for permission.

Tripping Over Engineering: Going Nuclear

LGM-25C Titan II Complex
Sahuarita, AZ [TMM]
135 built, 54 silos
Active 1962-2003
Largest US ICBM, highest single warhead yield US ICBM

Titan - (Greek) Second generation of deities who revolted against the parent deities.

Titan II launch {NMUSAF}.

Tripping Over Engineering: Going Nuclear

No Lone Zone, Two Man Policy Mandatory. Practically all areas with contact or influence on the security and functionality of nuclear weapon systems. At all times requiring at least two men, competent in the related system function, maintaining full observation of each other. Non compliance would be, at a minimum, unpleasant.

The Titan I only had a three year service life, but it was functionally a backup to the Atlas program and a developmental step to heavier lift. While the Titan I was still LOx/RP-1, it was the first true multistage ICBM. The Titan I went directly into a hardened silo complex. Like Atlas F, it was fueled in silo, boiling off oxygen, with the elevator lifting a fully fueled missile and a massive blast deflector to the surface. It managed to achieve this process in fifteen minutes. It was peculiar in being deployed in a three pack configuration, the only move away from individual distributed ICBM launchers. There was always the concept that each missile had specific blast protection, but any blocking or racking that prevented full opening of the silo door and the missile was trapped. If you packed the vent system, how could you even load the LOx. The Titan I also had particularly limited guidance with inertial providing autopilot control at launch, precision outbound targeting by radar and radio command. With the radar station only tracking and directing one missile at a time, the control could be killed by taking out the antennas, control center or a nuclear strike saturating RF. When the Titan II replaced the Titan I, there was a glut of retired Atlas available already with satellite launch experience. Out of 101 produced, most were scrapped and there are a few remaining on display. [NMUSAF] has Titan I and II standing side by side with other ICBMs to show the glory of scale.

A critical plan was to have a missile that could be fully fueled and ready to fire. It was expected that solid rockets would have an extensive development period before they could achieve adequate shut off control to properly target an ICBM, continuing a necessity of liquid fueling on Titan II. It also achieved two of the great cold war titles, the largest yield single weapon delivered by any US missile and the largest US ICBM. This punch had a purpose, the intent to crush hardened Soviet ICBM silos.

Tripping Over Engineering: Going Nuclear

Besides the two stage configuration, Titan I had another development to give the Titan II, the base Aerojet LR-87 first stage and LR-91 second stage. LR-87 was one of the first rockets to be readily adaptable to three different fuel/oxidizer combinations.

The fuel combination selected was nitrogen tetroxide with Aerozine 50. Aerozine 50 was equal weight hydrazine and unsymmetrical dimethylhydrazine lowering the freeze point and reducing spontaneous decomposition better than either fuel alone. While these remained liquid at ambient temperatures, they have maximum temperature limits to remain properly stable. In addition, both materials were corrosive (and destructive to skin on contact with liquid or vapor), toxic (inhaling of vapors or reactive products), hypergolic (instantly reactive in contact with each other which makes launch ignition very simple, also when not in launch sequence and when in contact with a range of other chemistries). Fueling required very specialized systems and encapsulated suits for crews with temperature control for propellants and the missile itself.

Each of the two main engines had an automated butterfly valve for fuel and oxidizer. These provided for the actual control of the engines. It was later determined that the ready launch capability should have more security in launch permission since there was no intervention after ignition. An actuated butterfly valve on the first stage oxidizer line was linked to a coded control panel in launch control to subdue armageddon. It was preset and required a specific code to open with only a three code fail attempt lockout. The BVLC butterfly valve lock code was sent with the SAC war orders. This was a major security upgrade and preempts later PAL systems. The additional valve also provided an additional block in the event of valve weepage while stored. {Schlosser 2013}

The Titan speed to launch was greatly improved by in silo launch, shooting back the 740 ton door and lighting the torch. Unfortunately, all of those hot high velocity combustion products have to go somewhere, and preferably not around the rocket. A set of diverters and ducts divert the gases to

Tripping Over Engineering: Going Nuclear

the surface and away from the silo mouth. The firing of the rocket in the silo functions as an eductor pushing all gases down and away from the missile, but blowing enough gas to generate 430,000 lbs of thrust generates enough acoustic pressure to destroy the missile. Directly under the engines are four venturi nozzles shooting a total of 9,000 GPM of water at a converging point. The water and flashing of steam provided adequate acoustic attenuation during the launch. An awful lot to ask of a little water.

Titan II launched the Gemini series of spacecraft. Expansions on the design as Titan III & IV were the core of US Air Force space launching. This continued with the hypergolic fuels in spite of the handling precautions required. Solid rocket boosters provided extensive capacity upgrades. They were finally retired due to competition from cheaper commercial launch vehicles. Considering that there were only 54 Titan IIs deployed on missile alert, the entire family of Titans, including extensive Titan I testing and development, achieved 322 successful launches out of 368 firings. Even Cassini-Huygens got its ride on a Titan IVB/Centaur.

9 Aug 1965, Launch Complex 373-4 near Searcy AK was being hardened. Warhead and RV were removed but missile S/N 62-0006 was full of fuel and oxidizer. A fire broke out during hot work and only two of 55 workers survived. This missile was overhauled and eventually loaded at 374-7 near Little Rock.

Following a fueling exercise, it often took weeks and three or four chargings until pressures stabilized. While crews were adding nitrogen to an oxidizer tank, a nine pound socket was dropped (begging the question of a basic gas maintenance issue that requires a nine pound socket). Typically, dropped tools would fall fairly directly into the exhaust deflector pit, with the crawling of the pit to retrieve the tool as adequate penance for the offender. This time it managed to bounce off a thrust mount and punch the fuel tank, spraying fuel into the silo. Because of the fuel vapors, it was hard to visually confirm a fire and the fire detectors were not triggered. But the oxidizer pressure was falling and fuel pressure was rising. With the missile loaded for launch, too low tank pressure and it would buckle and crash in on itself. Over pressure and it

Tripping Over Engineering: Going Nuclear

would blow out, causing a range of damage before once again collapsing on itself. Either would result in hypergolic oxidizer and fuel from all stages mixing in a perfectly configured concrete combustion chamber. S/N 62-0006 exploded the next day, lifting the 740 ton armored cap off the silo and shooting the 9 MtE warhead 200 yards. The final nail in the Titan II coffin was 18 September 1987 at Missile complex 374-7 near Little Rock AK. "Command and Control" provides an excellent detailing of the incident {Schlosser 2013}.

The [TMM] South of Tucson has a full silo and launch complex with motors, warheads and fueling equipment open for tours. The complex even had its opportunity for movie fame, In Star Trek: First Contact, Zefram Cochrane developed the first warp drive and installed it on the Phoenix, looking much like a Titan II in the silo.

It isn't designed and built as a modern disposable missile. It looks like a proper aircraft including bare formed and riveted metal. Sort of like a space age Lockheed Constellation.
There is a Titan II standing at the [EASM] with decks to allow more complete viewing. There is a split Titan II at [NMNSH] where you can stand between the stages.

Tripping Over Engineering: Going Nuclear

LGM-30 Minuteman
MMI - 800 missiles - 1962 to 1972
MMII - 500 missiles - 1966 to 1995
MMIII - 550 missiles - 1970 through 2020?
1st storable ready to fire ICBM
1st solid fuel ICBM
1st in silo launched ICBM

Minuteman - American militia who were ready for service at a minutes notice.

Atlas and Titan were similar to aircraft, exacting extensive interaction in maintenance and fueling. Solid fuel meant that Minuteman was always on ready standby and crews didn't have the same level of contact with the missile. The networked launch centers were remote from the dispersed silos without the ownership of a particular missile. Even maintenance was more modular with fewer fixing issues {Dobbs 2008}.

During the Cuban Missile Crisis there was a rush to commission MMI, the first solid fueled ICBM. Electrical issues were a problem in launching with many cases where the two key system was bypassed into a single key launch. To maintain some level of launch security, the silo door charges were disconnected with maintenance crews on standby to reconnect. Nuclear security was deferred to enlisted crews sitting out on the northern plains in the approaching winter, knowing they would either be hit by inbound or possibly the blast of outbound {Dobbs 2008}.

MMII increased throw weight, reliability and accuracy with greatly modernized electronics. A revolutionary change was a liquid injection vector system on the second stage nozzle eliminating the complexity of gimballing on a solid rocket nozzle. The reentry vehicle had specific features to counter ABM systems.

MMIII introduced MIRV permitting a range of terminal guidance on three smaller more accurate warheads. Third stage received similar control to MMII liquid vector second stage. The MIRV package further enhanced decoy deployment to distract ABM systems.

Four test runs in 1959-1960 attempted to reduce the fixed target exposure of silos. Rail box cars made by St.Louis Refrigerator Car Company were loaded with MMI missiles to wander the US rail system. This prototype missile train demonstrated the mobile capabilities and was considered for later ICBM systems. A rail car prototype is at [NMUSAF].

The ultimate mobile launcher would be a non strategic aircraft launching an ICBM. Fall 1974, a C-5A parachute skid extracted an 86,000 lb MMI. Nose parachutes stood the missile vertical and a shortened firing of the first stage demonstrated launch capability. Achieving the drop and launch with navigation synchronization pretty much indicates viability. It was canceled as being excessively problematic and expensive, but it took some nerve to do it the first time. To repeat with modern aircraft and missiles would just be an investment exercise. The actual MMI test C-5A is at the Air Mobility Command Museum, Dover AFB DE.

Tripping Over Engineering: Going Nuclear

LGM-30 Minuteman II Launch Control Center & Silo
Phillip SD   [MMNHS]

Minuteman combat crew {NMUSAF}.

With the Minuteman, the Air Force finally had the holy grail of storable ready to fire missiles.   While it did not have the warhead yield of the Titan, it was simpler and more compact. Improvements in communication and Permissive Action Links made it practical to disperse the launch centers from the missile silos, a far cry from Atlas/Titan individual silo launching.   Units of five Launch Control Centers (LCC) each in charge of ten silos, any one with the redundant capability to fire all 50 in the complex for a 1,000 missile fleet under 100 LCCs.

Tripping Over Engineering: Going Nuclear

158

MMII had a self-effacing institutional building with a compound fence and an array of antennas. No one would be seen entering the LCC since security and entry was through the building. The LCC had far fewer crew in the bunker maintaining security and support crew above ground.

The main point of the armored, buried LCC was survivability to achieve launching with some morale for the crew. Some manner of exiting was needed other than what would be a mess of collapsed and racked blast doors and shafts from a direct hit. In the upper end of the LCC away from the primary access was an escape tunnel. A hatch looking much like an oilfield clamp fitting leads to a culvert full of sand (to prevent blast collapse) rising to the surface.

In order to escape, the survivors unscrewed two bolts, dropping the clamp and hatch cover to swing around on chains while they were just trying to hang on. Sand would pour into the LCC, survivors crawling up the culvert while pushing the sand down. Ascending from the dirt to the warm radiation and breathing deeply of the fresh fall out.

Compared to the direct control and intensive maintenance design of large Atlas and Titan sites, the MMII silos were sparse with power and phone lines, some very secure antennas, and a big herky slab of a silo door. There were no buildings or personnel inside the barbed wire topped motion detecting perimeter fence. If someone tried to break into the site for nefarious purposes, security would sprint from the launch center with orders to use deadly force if necessary. In the days before convenient closed circuit television, they would also sprint to the site with orders to use deadly force every time a pronghorn decided the fence was the best place to scratch off some pesky ticks.

Tripping Over Engineering: Going Nuclear

LGM-30 Minuteman II Training Silo
Ellsworth AFB [SDASM]

Transporter Erector at training silo Vandenberg AFB {NMUSAF}.

Tripping Over Engineering: Going Nuclear

160

The isolated Minuteman silos feel much different from the prior integrated silos with launch centers. A cap and a combination lock lends admission to the Personnel Access Hatch. Steps down to the launch control room and inner sanctum of the silo. The proximity of the huge electronics racks to the silo indicates an acceptance that any launch of the missile is a one shot deal. The MMII is petite compared to the Titan II, the painted solid rockets appear practically sterile.

In order for inertial navigation to find its way, it has to know where it starts. Since a silo tends to be quite stable in its location, the geographic launch position is a simple input. But, to keep it from going the wrong way around the world, it also demands an accurate azimuth reference. Typically, the farther away the reference, the finer the direction is datumed. One was initially chosen as far away as possible with the most precise location. A port on the silo viewed Polaris, a sighting would dial in true north. Unfortunately, this did not work well in regular operation and a set of datum monuments were set out around the silos. A tedious process of shooting two theodolites would provide the necessary datum. It would have been romantic to stay with the ancient mariner and steer by the light of the north star, shooting theodolites definitely lacks equivalent poetry.

Adjacent sits the elevatable rocket transporter which would simply slide the missile in or out of the silo. A special truck transports the aero package for installation after the rocket is positioned. This truck transports all onboard guidance technology and the warhead with special climate control and extensive security. Maximum protection of the package in handling was provided by parking the truck over the silo so the process was entirely indoors.

There is a nice aircraft museum outside the base where you catch the silo tour. In order to get to the silo, there is a nice tour of a modern B-1B base. A Minuteman launch procedures trainer is at [NMUSAF].

Tripping Over Engineering: Going Nuclear

SURFACE-TO-AIR MISSILES

Ground defenses against aircraft were initially gun related. There continue to be rapid fire gun systems for close in air defense. In the early Cold War the Army continued to use anti aircraft artillery for city defense. The limit of artillery was always range, even if you can reach altitude, a significant ring of installations is required around a major city. Higher and faster aircraft finally exceeding gun capability.

With the advent of guided missiles, longer range weapons could be fielded out of fewer sites. Range and speed was readily achieved with improving propulsion, but accuracy in early systems had its limits, particularly if the weapon had to target an individual bomber in the cloudy radar return of a formation. Nuclear warheads provided a destruction level to offset the imprecision.

With the strategic focus moving from bombers to ICBMs, systems improved from anti-aircraft to anti-ballistic missile. Engaging higher and faster bombers lead to engaging much higher and faster ballistic missiles. The effectiveness of anti-aircraft missiles altered bomber and missile tactics. Bombing changed from attacking at the highest possible altitude to the lowest possible altitude. This encouraged a return of smaller cruise missiles for low observable, low altitude penetration.

Tripping Over Engineering: Going Nuclear

MIM-14 Nike-Hercules
145 Army batteries
Service 1958-1975
SF88 Golden Gate NRA [NMS]
1st nuclear antiaircraft missile

Nike-Hercules - (Greek) winged goddess of victory + religious hero renowned for great strength and performing the twelve labors of Hera.

NIKE HERCULES TRANSPORTER - LAUNCHER (GOER) PROTOTYPE

While almost extensively a fixed base missile, a portable prototype on GOER, one of the worst riding and handling Army vehicles {RA}.

The MIM-3 Nike-Ajax became the first operational guided surface-to-air missile with a conventional warhead, solid rocket booster and liquid fueled sustainer. The limitations in navigation and guidance instigated investigation of replacing the warhead with the W9 from Atomic Annie, but sometimes bigger is most certainly, better.

A shape ubiquitous to Cold War science fiction, the doomsday rocket. The Nike-Hercules missile; 41 feet tall, 10,710 lbs, 150,000 ft ceiling, 90 mile range, >Mach 3.65, and a 20 KtE nuclear warhead. 145 missile batteries were operated with

Tripping Over Engineering: Going Nuclear

~25,000 missiles built.   Major cities that did not have a Strategic Air Command missile or bomber presence were still potentially ringed with nuclear missiles (each equaling the yield of Fat Man) on a 36 second track to launch time.

The W7 or W31 warhead allowed a single missile to get "close enough" or potentially engage multiple high altitude bombers in formation.   Unlike a conventional anti-aircraft warhead which has a kill defined by blast and projectiles, the airborne nuke has blast, heat, flash and radiation.   Without physically knocking the aircraft out of the air, potentially blinding crews or killing electronics to prevent targeting and detonation.   Nike Hercules was built specifically for nuclear capability with a range of yields, but conventional warhead options were maintained.

Growing from Ajax to Hercules, it was scaled up for range with all solid fuel rockets.   Nike Hercules was replaced by Patriot, but had one distinct progeny.   The Nike Zeus was Hercules on steroids to intercept ballistic missiles up to 170 miles ALTITUDE.   Nike Hercules launched the last US atmospheric nuclear test Shot Fishbowl Tightrope on 4 November 1962 specific to high altitude missile defense development.

Initial control ran analog computers tied to long range radar with local acquisition, tracking and ranging radars.   Later systems provided digital control with multi frequency radars in an effort to manage electronic countermeasures.   Lacking internal control, the missile was entirely ground controlled; launched on a computed intercept, en route corrections transmitted and a signal for detonation.   There was enough automation to manage launches so that the fail of a missile to launch would quickly fire the next on standby.   As an Army system, a surface-to-surface function with a barometric trigger gave a ground engagement or even a coastal artillery capability.   At maximum range, a ballistic anti-aircraft missile.

In the Marin Headlands there is an area with various old military facilities and artillery batteries from generations of protecting of the bay.   Get there at the right time and you may see the winged goddess of victory and the god of strength standing strong.   On my tour, the missile was not

Tripping Over Engineering: Going Nuclear

erect, but we rode the elevator down into the hangar while touching the missile. The lack of installation hardening clearly indicates that defense was a one shot opportunity. If the bombers weren't stopped, there was little reason to protect the base.

Near Homestead FL, HM69 is under NPS management. It is less intact than SF88, but interesting as a later incarnation of site design. There is an unattended site near Roswell NM which was never operational, but somewhat different in design. W-50 is on an old AAF airfield, a runway as a road, with the launch shelters still standing. Watch for rattle snakes and owls, I found both.

[NMNSH] has a Hercules on a fixed base launcher. [WSMR] has a Hercules and an Ajax on transportable launchers. {Lonnquest 1996}

CIM-10 BOMARC
700 missiles both models
In service 1959-1972
Only Air Force surface-to-air missile
BOMARC - Boeing + Michigan Aeronautical Research Center

Four BOMARCs in ready position over coffin launchers.
{NMUSAF}.

BOMARC was originally a pilotless fighter as the XF-99, technically making it the last pre-century fighter. The Army was in charge for surface-to-air missile systems, so the Air Force kept control of the project purely through nomenclature.

47 ft long and 15,500 lbs with a W40 warhead lethal to a bomber within a 3,000 ft radius. A pair of ramjets underneath, requiring a boost rocket in the same manner as Navaho. But unlike most other vertical and ramp launched air breathers, the boost rocket was internal to the main fuselage. A major change to B model was a solid rocket booster with 440 mile range.

The A model had a 250 mile range with a liquid fueled booster. Hydrazine and red fuming nitric acid was toxic to breathe, nasty to handle, reactive or corrosive to practically anything it touches and not loaded until launch commitment. The movie version of fueling would show the harried control

Tripping Over Engineering: Going Nuclear

crews watching radar with staccato communications as the fuelers watch the flow meters turn for two full minutes with the Muzak of "Take Five" playing softly in the background.

After the wait, off it rips. In the B, up to Mach 3 and 100,000 ft, the highest intercept at the time. The Semi Automatic Ground Environment (SAGE) integrated radar systems provided for the advance fueling, launch and initial guidance from the 14 missile sites with onboard radar guiding the last ten miles. Terminal radar on the B was the first production use of pulse doppler. While the control system of the A was still vacuum tube, the B received a solid state upgrade. The use of all flying tail surfaces is typical on supersonic aircraft, but ailerons rarely pivot the entire wing tip. The Hjelmslev quadrilateral surfaces also give it the pernicious character of a battle axe.

Even though entirely engineered for high altitude intercept, there was an aspiration for low altitude engagement following the shift in bomber tactics. Line of sight on coastal radar could only reach 37 miles at 500 ft altitude with the short contact limiting intercept vector data. This was addressed on the B through a "Pattern Patrol" intercept. Multiple missiles would launch to individual area allocations and hunt on terminal radar, an early step toward modern autonomous weapons.

The BOMARC was adopted by the Canadians at the same time they killed the Avro Arrow, a large high performance interceptor. In some ways considered to be the death of the Canadian military aircraft industry, BOMARC indirectly destroying planes. Many were expended as drones on retirement giving a true high speed high altitude intercept target.

Apparently some have been pulled from less protected exhibits due to Mag-Thor, a magnesium thorium alloy that has both excellent high temperature mechanical characteristics and some pesky persistent radioactivity. BOMARCs advertised at [MOA] & [AFAM] were not at expected locations due to this, so if you are looking specifically for a BOMARC, confirm with the location. Currently known to be at [NMUSAF] & [NMNSH].

Tripping Over Engineering: Going Nuclear

AIR-TO-AIR MISSILES

Air-to-air combat in WWII experimented with other weapons systems, but guns continued to own the sky. Through the 50s and 60s guns were removed from fighter designs only to be added later. The new generation of fighters still have guns.

The significance of missiles and guidance was to achieve a standoff from the target and provide a more independent and multiple kill capability. Where surface-to-air systems had the space of fixed base operations for control, vacuum tube analog systems were limited when stuffed into an aircraft.

If guidance technology can only get you so close, just make a bigger blast.

Tripping Over Engineering: Going Nuclear

AIR-2 Genie
Operational 1957-1984
Over 5,500 produced.
19 July 1957 Plumbbob John nuclear test
1st and last air-to-air nuclear rocket.
Genie - A spirit capable of granting wishes when summoned.

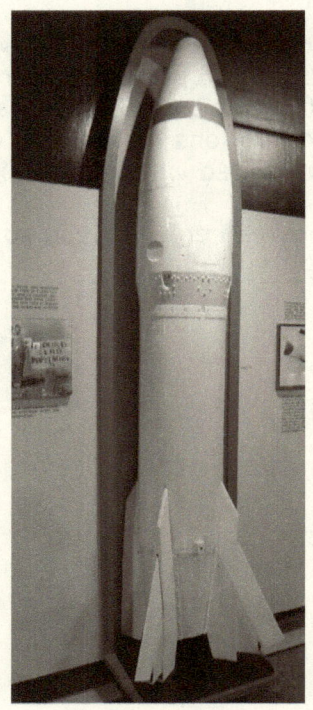

Genie at [AFAM] with fully extended fins, up close and personal {LB}.

With a few exceptions, immediately after WWII, most fighter aircraft still had a particular dependence on machine gun and cannon systems. As was known from heavy bomber attacks into Germany, huge formations can pretty much overwhelm defenses. Unfortunately for bomber crews, this overwhelming method also depends on significant attrition for those that are absorbing the defensive attacks. Immediately after WWII, the push started for air-to-air guided missiles. The AIM-4 Falcon was test fired in 1949, entering service in

Tripping Over Engineering: Going Nuclear

1956. But there was also a parallel air-to-air missile system that was developed based on an entirely different approach. At 822 pounds and 9 ft 8 in, it looks less like a rocket and more like a 1000 pound general purpose bomb, all it needs is a coat of olive drab.

Unguided with a 1.5 KtE warhead and a 1,000 ft kill radius. The W25 warhead contained the first US sealed pit design, fully encapsulating the nuclear material, but also preventing making the pit safe. All safing had to be in the ignition and explosive systems. With a time delay detonation, the aircraft fire control system managed the targeting and provided adequate aircraft separation. Mach 3.3 and a 6 mile range for firing to detonation under 5 seconds. This is also happening while the aircraft is headed directly toward the target at about 5 seconds per mile. While the best guided missiles are referred to as fire and forget, this would probably be more focused on fire and RUN LIKE HELL.

The Genie was the only air-to-air nuclear weapon test. 19 July 1957 Shot Plumbbob John was fired by an F-89J over Yucca Flats at 20,000 feet. A 2 KtE yield warhead was used as five officers and the cameraman stood directly under the detonation to demonstrate that there were few ground hazards. The actual test F-89J is at [GTF]. The actual missile is not available for viewing, it was, well, vaporized.

The Genie had a pretty significant service life from 1957-1984. It was around well after far better missile systems were in service. Besides the F-89, it was carried by F-101Bs and F-106As configured as air defense interceptors. Peak deployment in 1958 stocked 367 ready missiles in 42 fighter squadrons {Polmar 2009}. It was also used on Canadian F-101Bs with dual key security under US custody. Improved guidance capability eventually superseded the method of massive warheads.

While Slim Pickens rides an unspecified yet scary H-Bomb off the B-52 in Dr. Strangelove {Kubrick 1964}, Genie looks more like something at the supermarket that you would pay a quarter for your kids to ride. There are a number of Genies on display, some are at [NATM], [AFAM], [NMNSH], [WSMR] & [MOA].

Tripping Over Engineering: Going Nuclear

F-89 Scorpion
19 July 1957 Plumbbob John nuclear test
1st guided missile fighter
1st nuclear missile fighter
Operational 1950-1969
1,050 Aircraft

Scorpion - A terrestrial arachnid with a venomous stinger on the tip of its tail which it holds curved over its back.

YF-89A (46-679), gun nose, minimum tanks, young and lean before entering service, but always with those tractor wheels {NMUSAF}.

In the same manner that bombs and bombers are in the same discussion, air-to-air weapons have no meaning without fighters. The preliminary specification started 28 August 1945 as a high speed radar equipped interceptor with performance requirements ensuring a jet design. It was to be twin engined with four forward and two rearward guns including targeting to 15° off centerline. A swept wing was considered, but the straight wing was kept for lower speed control. Since it was somewhat considered a replacement for the P-61, the WWII practice of night intercept was to set up on the target and calmly stalk into a quick clean kill position. There was no dogfighting while following a voice on the radio to intercept, then creeping up with smudged blobs and lines on a CRT to a firing position.

There were numerous revisions of the XP-89 before production started flying as the F-89A under the revised

Tripping Over Engineering: Going Nuclear

nomenclature in September 1950; pilot and radar operator seated in tandem under a bubble canopy with an AN/APG-33 radar, six fixed forward firing 20mm cannons, twin afterburning J35s, fixed tip tanks, 16 underwing hardpoints, hydraulically boosted flight controls and split ailerons.

The F-89B/C had a range of structural and engine issues which were fixed for the F-89D to enter service in 1954. The D had the cannon removed from the nose with improved radar and intercept computers and additional fuel forward. Instead of guns, the wingtip pods were configured to carry 104 2.75 inch FFAR (Folding Fin Aerial Rockets), known more affectionately as "Mighty Mouse". It turned the F-89 into a giant flying double barrelled shotgun. At least compared to the fuselage mounted FFAR fighters, the F-89 crew was not enveloped in smoke during the firing.

There were questions of intercept effectiveness when F-89s were scrambled to shoot down an aircraft in the "Battle of Palmdale". An F6F target drone from Point Mugu lost control, wandering off range. F-89s were scrambled to down it over open country. The automatic fire control system failed to achieve lock, forcing manual firing. The installation of the fire control system removed the original gun sight, so they dialed up the salvo rate and sighted with the Mk-1 eyeball. Two aircraft shot full loads totaling 208 Mighty Mice, starting brushfires around the area as the F6F ran out of gas and gracefully glided to earth until mowing a power line and cartwheeling to disintegration.

The F-89E/F/G were used as development models for a range of improvements. The F-89H entering service in 1956 once again with upgraded fire control and an entirely different weapons package. Massive wingtip pods provided internal pop out mounting of six Falcon missiles comprised of a mix of GAR-1/2 for radar or infrared homing and 42 FFARs. The final F-89J replaced the wingtip pods with 600 gallon tanks and two Genies mounted on underwing pylons. Four additional pylons could be added for additional Falcon missiles.

The H model became the first US aircraft with guided air-to-air missiles. The J model became the first US aircraft with

Tripping Over Engineering: Going Nuclear

nuclear air-to-air weapons and the only US aircraft to do a live fire of the Genie or any other air-to-air nuclear weapon.

The straight wing behaved well from top speed down to stall characteristics with civilized stability for long flights and intercepts. All controls were hydraulically boosted to stay pleasantly light throughout the flight envelope. This kept it relatively aerobatic and the split ailerons provided plenty of braking for descent maneuvering. While there was an afterburner, the earlier engines did not have the same kick in the tail as a modern engine. There was a comment that afterburner on takeoff was a pop and a nudge with notable runway consumption. It was one of the last fighters without aerial refueling, but 2,365 gallons on the J model gave reasonable range and endurance. Besides, the intent of an interceptor was to get airborne hit the bombers and return home. {Higham 1975}

Jet aircraft development through this period was rampant with new or revised models almost monthly. The F-89 was outstripped in performance by the supersonic interceptors with more integrated radar and tracking systems. The Mighty Mouse was kept for a number of years evolving into the current Hydra 70 rocket. The GAR-1/2 Falcon missiles were a stepping stone for further radar and infrared guided missiles.

A lot of fighters get a range of other duties and deployments over their lives, but the F-89 lived out its life as an air defense interceptor well after being outperformed by other fighters. There are a large number of F-89s on display. One in particular, 53-2547 is on display at [GTF] in a proper F-89J configuration with a pair of Genies mounted. This is the only aircraft to have fired and nuclear detonated a live Genie in the Plumbob John test shot. F-89s with Genies mounted are also at [AFAM] & [CAM].

Tripping Over Engineering: Going Nuclear

AIM-26A Nuclear Falcon
Service 1961-1971
About 4,000 total nuclear and conventional
1st & last guided nuclear air-to-air missile.

Falcon - Bird of prey genus Falco.   alternate, small light   15-17th century cannon.

AIM-26A and AIM-4 on F-102A internal popout mounts, at [MOA] {LB}.

The AIM-4 was the first Air Force guided air-to-air missile, but the guidance of the time was debatable. Not to miss out on a combination of the best features, the guidance capability of the AIM-4 and the W54 compact warhead from the Davy Crockett program became the AIM-26A in a distinctly enlarged missile. According to a Hill AFB Fact Sheet, between radar guidance and the nuclear warhead, the most powerful air-to-air missile ever deployed. The risk of the nuclear warhead for low altitude aircraft led to the conventional AIM-26B which armed Swedish planes well into the 1990s. Apparently range from nuclear detonation for the aircraft was not considered to

Tripping Over Engineering: Going Nuclear

be an issue with the AIM-4, AIR-2 and AIM-26 all having a six mile range.

The AIM-26A was only deployed with the F-102, a high performance all weather interceptor built around the fire control system. In an effort to maintain the supersonic capability, weapons were internally stowed. It originally carried six AIM-4s and still two dozen Mighty Mouse rockets with one AIM-26A and three AIM-4s on later loadings. Apparently one big boom beats shotgunning with little rockets.

After retirement, the nuclear warheads were installed in the tactical AGM-62 Walleye TV guided bomb (trainer at [NMNSH]). This was the first weapon with a fire and forget vision tracking. Prior TV guidance required a command director to watch the screen and fly the weapon all the way to impact. The larger conventional Walleye II was used effectively on precision targets in Vietnam and Desert Storm.

Nuclear Falcons on display at [NNAM] & [MOA] has one mounted on an F-102A.

Tripping Over Engineering: Going Nuclear

BIG GUNS

With improved air delivery, missile capability and guidance, jobs within the range and shell capacity of artillery are still often best engaged with artillery. Within its limits, it is an efficient, effective and accurate weapon system.

While artillery was once all weapons beyond the infantryman, weapons systems have reduced the artillery fielded. The Army used to be everything from 75mm to 240mm as field capable, it is now 155mm as standard with a lightweight 105mm. The larger and smaller artillery functions have become rocket and missile applications. But the narrowed range of artillery also has a greatly expanded capability with rocket assist, in flight shell guidance, and the benefits of modern digital networking and control. A single artillery piece firing multiple rounds with simultaneous arrival continues to be a creature of wonder.

Artillery particularly benefited from the miniaturization of nuclear devices. But, there are stringent constraints in artillery; acceleration, rifling spin and critically, still fit in the barrel. This is particularly problematic with all of the electrical systems required for the early implosion devices. The use of a gun type warhead to simplify and reduce the cross section moves the other direction. To what extent can firing movements risk detonation or a fizzle. For the ultimate nuclear weapons engineering, it is hard to perceive of tougher design requirements than artillery.

Tripping Over Engineering: Going Nuclear

M65 Atomic Annie
Service 1953 to 1963
25 May 1953, Upshot-Knothole Grable nuclear test
20 guns deployed
1st artillery fired nuclear weapon

Upshot Knothole Grable {LANL}.   The classic shot.

On 25 May 1953 was the only nuclear weapon test fired from the barrel of artillery.   The detonation was only 7 miles from the gun on Frenchman Flat at [NNSS].   The seductive military designation for the test was Series Upshot-Knothole, Shot Grable.   The Army plan for the nuclear battlefield simply required upgrading to nuclear projectiles, not that nuclear capability required an inherently different weapon mentality.

Development started with the largest mobile gun in Army service, the 240mm M1 Howitzer often referred to as the "Black Dragon".   Its mobility required artillery tractors

Tripping Over Engineering: Going Nuclear

hauling the barrel unit and the carriage unit. A crane would travel with the gun to dig out the recoil pit, place the 20 ton carriage and mount 12.5 ton barrel. It was claimed that this could "race" (at 21 mph) to its location and "immediately" (1.5 hours) set up for firing {Lodge 1945}. Of course this was much more mobile than the WWI predecessor M1918 240mm howitzer which transported in four units with WWI era equipment. Strangely enough, it was a copy of a German PreWWI 280mm Schneider (best described as a siege gun) with changes for more modern 240mm ballistics. The Black Dragon was simultaneously developed with the 8 inch Gun M1 (Army longest range field artillery of WWII). Both of these monsters were researched as self propelled guns which would have still had very limited mobility. The 240mm was used against hardened targets in WWII and Korea (some remain in Taiwan as shore guns).

In 1949 the Army started developing an artillery piece that could fire a nuclear weapon due to free flight artillery rockets lacking the accuracy of a gun. By the time the M65 was fielded in 1953 it weighed 83.8 tons with its two specialty tractors. The carriage section was related to rail gun designs with two tow tractors using drop neck connections to carry the gun much like railcar trucks. As a homage to the big German rail guns, it was typically referred to as Atomic Annie. One hour from transport to firing including nuclear assembly was a quick set up, particularly with its size. This gave it more shoot and scoot capability, but not enough to overcome the 84 ft long 16 ft wide mobility limits.

In order to combine vehicular mobility with the projectile and warhead requirements, the 280mm bore was odd for the fleet, but it gave the minimum size gun to achieve the requirements. The Soviet response was to build four 16 inch self propelled howitzers. The short service life indicated that it was even less useful than Annie. Nuclear shells for the existing Army 8 inch gun were soon developed and eventually the even more common 155mm.

Typical of historic super guns, the capability never exceeded the inconvenience of the equipment and crew to move these behemoths. Missile capability would also quickly eclipse the performance. Modern land warfare already used significant

Tripping Over Engineering: Going Nuclear

aircraft with much greater delivery capability, eliminating the exposure and logistics of getting this beast within range of target. Even though it's field capability was limited, it was operated until 1963 primarily as a threat and prestige weapon.

The 43 foot barrel could huck the original 800 lb 15 KtE W9/T124 nuclear shell 15 miles with ~80 stockpiled between 1953 and 1957. The W9 was a gun type uranium device derived from the TX-8/W8 bunker busting warhead which was a derivative of the Mk-1. The bunker buster was configured to detonate after ground impact providing robustness for gun use {Polmar 2009}. Between 1957 and 1963 ~80 W19/T315 shells were improved W9 replacements at 600 lbs, 15-20 KtE and an 18.6 mile range. From May 1954 to May 1955 the W32/T332 240mm shell was researched and canceled. {Hansen}

Some of the W9 warheads were recycled into the T-4 Atomic Demolition Munition. This was one of the earlier "man portable" munitions meaning a crew of five loaded to capacity could carry it. There were also conventional shells for Annie. After retirement, there was a study on sabot firing them in the 16 inch Iowa class guns.

Only 20 of the guns were made, there are seven survivors. The original Grable test gun is at Fort Sill Oklahoma. The Virginia War Museum has the only 240mm prototype. A gun, shell and both tractors are at [NMNSH].

Tripping Over Engineering: Going Nuclear

M-28/29 Davy Crockett

17 July 1962, Sunbeam Little Feller I nuclear test
Last US above ground nuclear test
Service 1956-1971
400 nuclear projectiles produced
Smallest nuclear projectile weapon

Davy Crockett - American folk hero, soldier, politician and raconteur. Failed reelection to Congress for voting against Andrew Jackson's "Indian Removal Act". Killed at the Battle of the Alamo.

Davy Crockett light mount {NMUSA}.

Far less rare than Atomic Annie, with 2,100 weapons fielded between 1961 and 1971, far more ludicrous is the Davy Crockett M28 120mm & M29 155mm with M388 shell. Depending on the recoilless gun configuration, it had a max range of either 1.25 or 2.5 miles. The spigot launched projectile with a W54 warhead had a yield of 10 to 20 tE, no mega, no kilo, just tons.

Throughout the history of recoilless guns, the primary benefit

Tripping Over Engineering: Going Nuclear

of eliminating recoil is through equal momentum blowing out the breach as well as the muzzle, as dangerous behind as in front. The eleven inch diameter warhead precluded internal barrel firing typical of recoilless rifles. The result is ostensibly a fat finned mortar shell hanging out the end of the barrel on a launch spigot typical of old rifle launched grenades. Expecting the guns to be existing issue models, they were actually very specific designs for application. It was the first titanium gun in the Army, it had the first multizone recoilless system and one of the few recoilless spigot weapons. {AMC 1976}

The projectile has a safety system, so to speak, a switch for arm or safe, "is this thing on". Detonation altitude was simply selected with a Hi-Lo switch. Targeting was simplified with a coaxial 37mm spotter; center the crosshairs, shoot the spotter, if it pops on target, let fly the watermelon.

On 7 July 1962, Operation Sunbeam shot Little Feller II (at least Little Feller was actually descriptive) suspended the W54 warhead three feet off the ground with a yield of 22 tE. Little Feller I was fired on 17 July 1962 from the M29 to 1.7 miles with a yield of 18 tE, the last US surface nuclear test.

The size of the W54 warhead lead to its use in a couple of other applications. It was used in the Falcon air-to-air missile to make the AIM-26A, the US's only nuclear guided air-to-air missile. The light compact 51 lb warhead was the genesis of true backpack demolition nukes. While typically operated with motorized transport with the 500 lb M29 system, the smaller 170 lb M28 was intended as an INFANTRY weapon. Wiley Coyote couldn't have devised a more preposterous weapon. There is discussion that this was the device Heinlein considered when he put nuclear weapons on the infantry armored mecha suits in Starship Troopers. {Heinlein 1959}

There are several survivors including [NATM] & projectile only at [NMNSH].

Tripping Over Engineering: Going Nuclear

W23 Katie
16 inch Nuclear Naval Shell
50 Shells
Service 1956 to 1962
The only nuclear naval gun shell

Katie - There is speculation that this name started as shorthand KT similar to HE, HC or AP shells. Or fabricate a story of Katie being the name of an ex-girlfriend that rained destruction wherever she went.

USS Missouri (BB-63) salvo from turret #2, Korea Oct 1950 {NHHC}.

In WWII, only four of the originally planned six Iowa Class battleships were completed. They returned to service for Korea and Vietnam. The New Jersey, Missouri and Wisconsin fired their last combat shells during Desert Storm, the Iowa inoperable from a turret accident. Even into the aircraft and nuclear age, the Navy thought you could never go wrong with a really big gun. Few early missiles could achieve the accuracy of artillery. If you can already heave a big shell on target at a decent range, why not heave a nuke. Hence, the 16 inch Mk7 gun received 50 rounds of the W23 shell. Easy design; take Atomic Annie's W9 warhead, package as the standard 1,900 lb shell. A 16 inch nuclear naval shell that looks like, handles like and shoots like, a standard 16 inch high explosive naval shell. Out to 24 miles range, all you had

Tripping Over Engineering: Going Nuclear

to do was dial in standard ballistic settings and let it fly. Standard HEs even allow natural fall spotting.

Three Iowa Class ships were configured with segregated nuclear security magazines. Although the Iowa Class were the only battleships recommissioned postwar, little effort would be needed to fire Katie from US post WWI 16 inch battleship or shore guns. The Army had some interest for the shell since there remained 21 wartime 16 inch coastal guns. If nuclear coastal defense guns sound questionable, Camp Lejeune was once considered alongside Nevada for a possible nuclear test site.

Katie was in service from 1956 through 1962, a fairly short service life, but also a time of significant missile progress. Since the battleships fell in and out of favor post war, the Iowa class had three different periods of commissioning. The shells remained in service over several years while the ships were mothballed. It may have been more convenient to sabot fire the Atomic Annie shell from the 16 inch, since that was a consideration during Vietnam to consume 23,000 surplus Atomic Annie conventional shells.

I believe the only survivor is in [NMNSH]. At least one was specifically converted for Project Plowshares and there were a lot of Plowshares firings in a 15-20 KtE range. The W9 warheads had also been recycled to the T-4. Big artillery having been profoundly outstripped by aviation and missiles, warheads were removed for other uses and the shells simply scrapped.

Katie was destined for the Iowa class ships which were upgraded through the 1980s. The South Dakota and North Carolina classes were decommissioned post war so they are closer to the fire control configuration when the shells were inventoried. An optical rangefinder spanned the width of the turret. A 3,000 lb mechanical Mk-1 fire control computer corrected relative position, motion, wind and even the spin of the earth. Direct fire control operated on divide and conquer. One man cranking azimuth on the vertical crosshair and one man cranking elevation on the horizontal crosshair. Except for power assist, no different from the wartime 3"/50s. The only survivors of these classes, the North Carolina, Alabama and Massachusetts are currently on display.

Tripping Over Engineering: Going Nuclear

W33/W79 8 inch
8 inch Army nuclear shells
Service 1957 to 1992

M110 crew with M422 shell {LANL}.

Unlike Atomic Annie and Davy Crockett, the eight inch shells were intended for use in existing artillery. The M115 towed howitzer was a prewar design used from WWII through Vietnam with Cold War deployment in Europe specifically for the nuclear shell. Many were supplied as military aid to a range of countries, the final demise in those services simply from lack of ammunition. While there were some attempts and making the M115 self propelled, it lacked progress until the M110 was developed. This was the largest Army self propelled artillery used from Vietnam through the Gulf War. This had the benefits of the later rocket assisted shells. With improved 155mm and MLRS, these big heavy guns were limited in a more dynamic battlefield. Like the M115, multiple countries were supplied with the M110 and some still appear to be in service.

Tripping Over Engineering: Going Nuclear

The W33 was a gun-type warhead, with M422/T317 live and M424 spotter shells, 243 lbs with a 9.3 mile max range. Pit installed determined one of three yields, 1 KtE, 5-10 KtE or 40 KtE. Lacking ballistic equivalence to conventional rounds, special spotter rounds were required. 2,000 were produced between 1957 and 1965. While typically replaced by the W79, some may have been in inventory as late as 1992.

The W75 warhead was developed 1970-1973, but canceled due to cost and lack of performance improvement over W33.

The W79 was a linear implosion plutonium warhead with M753 live and M754 trainer shells, 215 lbs with a rocket assist for range to 18 miles. Three different yields were possible from 0.2 KtE to 2 KtE and W-79/0 warheads convertible to enhanced radiation by swapping a tritium module for the rocket. Ballistic equivalence allowed firing concurrent with conventional rounds as spotters. Permissive action links and a Command Disablement System to hard safe the warhead significantly increased security. 550 total were built 1981 to 1986 replacing the W33 with final retirement in 1992

Of the four gun-type fission weapons detonated by the US, two were W33 warheads. Shot Plumbbob Laplace detonated a 1 KtE warhead suspended at 750 ft by balloon on 8 September 1957. Shot Nougat Aardvark was a 40 KtE underground detonation on 12 May 1962. Both were performed at [NNSS].

Eight inch guns of various provenance are scattered around the US at AFL and VFW posts. Nuclear capability in such an ordinary piece of equipment. There is a W79 at [AMSE]. {Ramsbotham 1989} {Hansen} {Serchak 1980}

Tripping Over Engineering: Going Nuclear

W48/W82 155mm
155mm Army nuclear shells
Service 1963 to 1992
Smallest barrel fired nuclear shell

Designers with mock-up of W48 {BI}.

Like the 8 inch, the 155mm permits the use of standard firing systems. The 155mm is now the largest US gun with rockets and missiles replacing the larger weapons. The convergence to the 155mm as largest artillery is also sandwiched from below with the 105mm remaining as a light gun. The 105mm is declining for general purpose use with current development on a lightweight 155mm as a replacement. Current 155mm research has a 46 lb projectile with a range of 58 miles intended to kill ballistic missiles. Rocket assisted and guided warheads would allow an amazing range of options to resurrect the nuclear shell.

The 155mm nuclear warhead was an ongoing miniaturization

Tripping Over Engineering: Going Nuclear

from the 8 inch shell with linear implosion plutonium warheads.

The W48 warhead had M455 live and M454 trainer shells; 128 lbs, range to 8.7 miles and 100 tE. This round lacked ballistic equivalence to conventional rounds and required special spotters. 1,060 were produced between 1963 and 1969.

The W74 was developed 1969-1973, but canceled due to cost and lack of performance improvement over W48.

The W82 warhead in the M785 shell would have replaced W48 with a 1,000 scheduled, but was canceled due to cost (~10 times W48) and possible stockpile reduction. 95 lbs with rocket assist to 15 miles, less than 2 KtE with warheads convertible to enhanced radiation by swapping a tritium module for the rocket. Ballistic equivalence to standard rounds permitted concurrent firing with conventional rounds as spotters. Security was increased with PAL and a Command Disablement System.

Like the 8 inch gun, there are 155s sitting in front of VFWs & AFLs all over the US. These are impressive as an even simpler firing device. There is a W82 at [AMSE] and M455s at [NMNSH] & [NATM]. {Ramsbotham 1989} {Hansen} {Serchak 1980}

Tripping Over Engineering: Going Nuclear

POWER

Many early researchers in the Manhattan Project recognized the potential of nuclear power as a propulsion system for submarines. They did not know the program was weapons only until orders were issued stating that the ONLY work permitted shall result in the completion of the bomb {Rhodes 1986}.

The two great benefits for propulsion were the energy available on a single fueling and the freedom from combustion air requirements. But, that benefit also carries some fairly high costs restricting it to very limited applications. Before that was confirmed, there was a lot of experimentation.

Tripping Over Engineering: Going Nuclear

USS Nautilus (SSN-571)
Groton CT [SFM]
30 Sep 1954-1980
1st nuclear submarine

Nautilus - Captain Nemo's ship in Jules Verne's "20,000 Leagues Under the Sea". Power was implied to be something similar to nuclear.

Nautilus, initial sea trials 20 January 1955 {NAC}.

Electric Boat in Groton launched the first practical US submarine in 1899. On 21 January 1954 Groton launched another submarine revolution and the rules of submarining changed forever.

The USS Nautilus (SSN-571), the world's first operational nuclear submarine. While lacking the high speed subsurface hull of the Albacore, the GUPPY improvements of post war fleet boats were improvements over wartime hulls. Although sleeker than war time boats, the profile did not stand out significantly against the diesel-electrics of the time. Considered for production as a missile boat, Rickover did not want a failure of a missile system to damn the entire nuclear submarine program.

With air limits of propulsion eliminated, Nautilus set depth, speed and range records. New London CT to San Juan PR, 1,590 miles in 90 hours fully submerged for 17.6 mph average. While being able to chase or run at will, a specific hydrodynamic flaw was discovered. Above 18.4 mph, vortex shedding off the sail would approach the 3 hz hull frequency, generating cumulative damage and making the sonar useless

Tripping Over Engineering: Going Nuclear

over 9.2 mph {Polmar 2004}.

The first crossing of the north pole UNDER the ice cap on 3 August 1958 covered 1,830 miles submerged over four days {Polmar 2004}. This demonstrated the capability of the planned missile boats to disappear. The next duration limit was crew breathing air. In 1959 Nautilus was the first submarine to receive an electrolytic oxygen generator, with deployment now limited only by food capacity. Besides performing extensive research, Nautilus performed regular patrols, including October 1962 near Cuba.

Nautilus demonstrated the ability to achieve safe mobile nuclear plant operation. Other than some smaller coastal boats, the standard for long range fast attack and missile boats rapidly moved to nuclear power. It has also become the standard for US carriers for its ability to run unrefueled at high speeds and long ranges. The reactor on the Nautilus was much less problematic than that of the second US nuclear sub. The USS Seawolf (SSN-575) spent two years having its reactor replaced by a spare Nautilus reactor. But in 1958 the Seawolf did break the submerged record over 15,835 miles and 60 days with the original NaK reactor. The liquid NaK reactor on Seawolf had what is probably the coolest foible, the glowing blue halo of Cherenkov radiation along the sides of the boat {Polmar 2004}.

USS Enterprise (CVN-65)
Currently in limbo
Active 1961-2012
1st nuclear powered aircraft carrier
Longest serving US aircraft carrier
Enterprise - believed to be the name inspiration for NCC-1701, 2nd carrier and 8th Navy vessel named Enterprise

Navy's first three nuclear surface ships in 1964, B-T CVAN-65 Enterprise, CGN-9 Long Beach, DLGN-25 Bainbridge. Task Force One, Operation Sea Orbit. Around the world 30,565 miles in 65 days. On deck A-5(9), A-4(22), F-4(10), F-8(14), E-1(2) {NNAM}. 55 nuclear weapon capable aircraft.

Even during WWII, there was a regular concern on the limits of the existing fleet carriers. Postwar, there was a push for five new "supercarriers". On 18 April 1949 the keel was laid for the USS United States (CVA-58) and five days later the entire class was canceled, lost in the inter service fight for declining budgets. While the symmetric flush deck was ostensibly for operating strategic bombers, it would have provided a much more flexible capability. A 100,000 lb aircraft capacity allowing full deck rolling take offs with the

Tripping Over Engineering: Going Nuclear

most powerful catapults and arresting gear to date. This also dictated heavier aircraft and ordnance elevators to support the larger aircraft. A major killer of the flush deck was the process of ship and flight deck operation. It greatly reduced radio, radar and defensive gun placement options. Even given all of the technology, there are times when no vision system can replace stepping out on the bridge wing for an all encompassing view of the world.

Into the 1950s, improvements were added to remaining WWII Essex and Midway class carriers. They received angled flight decks so that a missed trap would permit an aircraft to go around or at least not pile into aircraft stacked for the forward catapults. This greatly improved launch and recovery rates. Arresting gear and catapults were also upgraded for heavier faster aircraft. These improvements would not be adequate for the aircraft that were to be developed. Of these carriers four Essex and one Midway class carrier remain on display.

The faster heavier aircraft incited a return of the supercarrier. Not to the extent of the United States, but a great improvement from the wartime ships. This would become the Forrestal, Kitty Hawk and John F. Kennedy classes. All aircraft related mechanicals were sized for modern jet aircraft. These were the first new builds with angled flight decks, steam catapults and optical landing systems, recognizable as modern CATOBAR carriers. These continued to be an oil burning ships, but deployment plan and aircraft fuel mixes allowed some flexibility in extending ship range. The John F. Kennedy is the last of these classes and is currently on hold to determine if a home may be found.

In the midst of the previous carriers was the first shot at the nuclear carrier. While the cost and operation of fewer large reactors would have been preferred, there was also the development period for larger reactors. Where carriers previously had eight oil burning boilers, Enterprise had eight reactors developed from the nuclear cruiser program. The most reactors installed on any nuclear vessel. This allowed quicker development with a less than optimal design. Fitted into an 1,123 ft hull, the longest combat vessel and longest carrier flight deck.

Tripping Over Engineering: Going Nuclear

Nuclear powered, but also known to be nuclear equipped. During the Cuban Missile Crisis on 27 Oct 1962 both Independence and Enterprise were within 150 miles of Guantanamo with forty tactical nuclear bombs on standby for loading on A4D Skyhawks {Dobbs 2008}.

Originally intended to be scrapped after reactor removal, it was placed in limbo for potential museum placement. Of all of the nuclear vessels used and retired by the US, the only other existing retired nuclear vessels are the USS Nautilus and the NS Savannah. Currently unavailable for visitation unless you have an inside connection with the Navy. But, Enterprise is one of a kind and we can hope for an adoption. {Friedman 1983}

NS Savannah

Only US commercial nuclear vessel
Launched 21 July 1959
Service 1962-1972
Preservation association http://www.ns-savannah.org/
Baltimore, MD under control of MARAD DOT

Savannah - SS Savannah built 1818, first steamship crossing of Atlantic 1819. Lack of commercial success led to pure sailing conversion. Wrecked 1821. Namesake for NS Savannah.

Savannah approaching San Francisco {NAC}.

Once again, I am discussing a vessel that has a very limited access. But exceptionally unique for the US. Even in the world, first of only four nuclear cargo ships. The only other one that was regularly operational was a Soviet ship that was linked with their nuclear icebreaker and arctic operations. And, unlike the others, an exceptional beauty of "Atomic Age" styling. No stacks or smoke, so sleek and clean. Old school cargo passenger combination stretched long and lean.

Savannah was part of Eisenhower's "Atoms for Peace" program. While nuclear projects like Plowshare raised a lot of questions where nuclear was heading, efforts in nuclear power, materials and medicine had positive results. But, it was a distinctive political project disconnected from the direction of the shipping industry. At the time, the most expensive cargo ship ever built, a high maintenance love affair.

Tripping Over Engineering: Going Nuclear

Babcock and Wilcox designed the PWR with their experience in commercial power, but outside of the reactor area are typical marine systems. Main propulsion was Delaval steam turbines. Sufficient diesel power was installed to run emergency cooling pumps and maintain steerageway. An early installation of dynamic roll stabilization gave a premier ride on a premier ship and also reduced sloshing conditions for reactor steam systems.

Operation required far more extensive controls and licensing to the point that it was just plain expensive to operate. The use of combination ships was ending in commercial operations. The Soviet nuclear icebreakers had specific benefits with nuclear power and have continued in useful service. It would appear that a modern compact reactor would make a modern ship safer and more cost effective, but the licensing and control overhead for these systems would always be problematic. Global distribution of cheap fuel oil was the final nail in the coffin, eclipsing nuclear operating cost.

Only carrying passengers for three years with freight capacity to low to be cost effective, it is estimated that 1.5M visitors toured the engineering space, claimed to be the most heavily visited nuclear facility in the world. While not there during my tenure, Savannah was serviced and stored several times near the Coast Guard base in Galveston TX. While military vessels are stripped of the reactors early in their decommissioning, Savannah was defueled with part of the radioactive systems removed, but the reactor is still installed.

Hopefully Savannah will get a permanent public home. If not, see if you can get a MARAD group tour. Don't miss seeing the high tech galley, with a new appliance called a "microwave oven". A Raytheon RadaRange weighing 750 lbs with water cooling for the magnetron. The leaders in military radar develop the world's fastest popcorn maker.

Tripping Over Engineering: Going Nuclear

HTRE-2/3
**Nuclear powered bomber program
Only nuclear jet engine prototypes
Testing 1956 to 1961
Idaho National Lab   [INL]**

HTRE - Heat Transfer Reactor Engine - colloquially as "Heater".

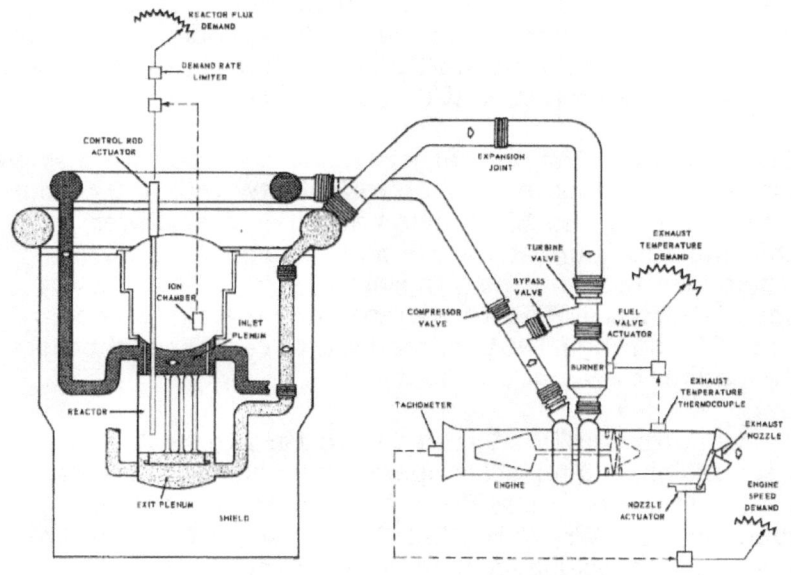

HTRE-1 Schematic   {Thornton 1963}.

In front of the EBR-1 building are three strange items. A yellow thing sitting on railroad tracks that looks nothing like a locomotive, it's a locomotive, with full radiation shielding. Two chemical plant looking devices sitting on railroad tracks that look nothing like jet engines, they're jet engines, nuclear jet engines. These were test beds for what was to power a nuclear jet powered bomber and a locomotive to move them.

If you intend to keep an aircraft on airborne alert for extended periods of time, what would be more practical from a fuel standpoint than nuclear.   Dispatch bombers for days, circling

Tripping Over Engineering: Going Nuclear

over open ocean, a veritable flying missile submarine.

A jet engine is pretty straight forward; compress some air up front, heat it up (normally with fire) spin the turbine to drive the compressor and shoot the rest of the energy out the back. Replace the fire with a high temperature air cooled reactor and you are on your way, except reactors tend to be heavy and the shielding tends to be heavier. Start with a J47 engine replacing the standard combustor with cold and hot side plenums through a bypass and external combustor. This allowed start and run of the J47 independent of reactor power for cooling during start and shutdown of the reactor or a range of system testing without going critical on the reactor.

HTRE-1 was the first nuclear jet engine test bed with a single turbine. Operating in 1956 up to 20.2 MW with 485.6 hours at over 200 KW and 150.8 hours at full nuclear power, achieving 100 hours at 1280°F and 44 hours at 1380°F turbine inlet temperature {Thornton 1963}. HTRE-2 was converted from HTRE-1 to operate more than 1,100 hours from 1957 through 1961, primarily with ceramic fuel pellets intended for the flight engine.

The twin turbine HTRE-3 tested from 1958 to 1961 was a closer design to a full scale operational engine. Besides providing for more mass flow, it had more aircraft type redundancies. Where HTRE-1/2 had a secondary compressor to provide for residual cooling in the event of loss of the turbine, HTRE-3 could provide for cooling and power at a reduced reactor rate if one jet went down. HTRE-3 even operated on a full nuclear only cycle with the engines spun and fired using only start heat from the reactor. A serious test considering the repercussions from a hung or hot start in this case is far more than just a burned combustor or gas generator wheel. It achieved 29 MW and 147 hours on full nuclear power operation {Thornton 1962}.

A B-36 was modified to the NB-36H with an onboard reactor to research the radioactivity issues for the crew. The reactor was operated but did not provide power, no other aircraft development was done. The test data generally indicated that a nuclear bomber was feasible, but it was one of those projects that would always have high accident consequences

Tripping Over Engineering: Going Nuclear

and high cost. Like so many strategic weapons projects, the growing ballistic missile capability became a much more reliable method.

While the tests ended with the HTRE-3, there had been development proceeding with twin and four turbine engine configurations. GE even built the mechanicals for a flight engine and tested performance by simulating the planned reactor thermal function. Even if reactor safety had been considered adequate, crew shielding would have always been problematic.

It is exciting enough near a thundering jet engine, it is overwhelming to picture these monsters howling away on nuclear. The GE APEX 900 series of declassified project reports give an amazing amount of detailed radiological instrumentation and metallurgical data, see GE-ANP on leehite.org.

Supersonic Low Altitude Missile (SLAM)
Project Pluto - Tory tests
14 May 1961 first run, 1 July 1964 canceled
Nuclear ramjet tests
Jackass Flats [NNSS]

Pluto - (Roman) God of the underworld and judge of the dead. Tory - American colonist supporting the British during the American Revolution.

Tory II-A, there are photos of C but the A was more imposing {WIKI}.

There was one project that made the nuclear powered bomber look like a perfectly intelligent aircraft, Project Pluto with the Supersonic Low Altitude Missile (SLAM). Strap some solid rocket boosters to a ramjet with a very high temperature reactor in the middle. Shoot it to self sustaining speed then it moseys on its way at Mach 3. As a cruise missile, it solved the personnel radioactivity issue by removing the personnel. Alert mode operated on radio command with attack on autonomous terrain matching radar at low altitude to deliver about sixteen bombs then dive into its final target. If it wasn't committed to attack, there was a consideration to put the missiles into a deep water trench of the Pacific. A graveyard at sea of smashed up live nuclear weapons and reactors. {Herken 1990}

Tripping Over Engineering: Going Nuclear

While the missile did not proceed beyond concept, there was engine testing at [NNSS]. Tory II-A had a few seconds of run time on 14 May 1961 and validated the progress to the next engine.

Tory II-C ran on 20 May 1964. It was found that a compressor adequate to drive the air supply would have been monstrous so there was a review of underground air storage, considering that there were nice nuclear caverns from subsurface testing around the area. The eventual system was 25 miles of oilfield casing manifolded to a rack of submarine compressors on loan from Groton. This blew through a four tanks containing over 1,000 tons of 1 inch steel balls preheated by oil burners. This provided the ramjet with inlet conditions of Mach 3 at 1,000 ft for 292 seconds, making 513 MW and over 35,000 lb thrust. The real magic in the engine was the same as on HTRE. While a combustion jet has issues with burn rate and contact temperatures in the engine, a lightweight nuclear has to have adequate corrosion protection of oxidation sensitive isotopes and adequate heat transfer rates below material temperature limits. The best route to maximum surface area, 500,000 individual fuel elements.

Cooler minds came to bear when the potential situation of a crashed SLAM was discussed. After ICBMs were developed with quick launch capability, SLAM was considered much less effective, just like the SNARK.

There are no parts to be viewed. The tour at [NNSS] doesn't go to Jackass Flats. The aircraft itself never went beyond concept. A logical progression of nuclear power and bomber mentality that considered the benefits and not the costs. Would the presence of this capability have reduced or aggravated the nuclear weapons race. There is a range of declassified documents for the program on {DTIC}. Even as a concept, this is an idea to wild to ignore.

Tripping Over Engineering: Going Nuclear

NERVA
Only nuclear rocket prototypes
NERVA - Nuclear Engine for Rocket Vehicle Application.
Nerva - Roman emperor for 15 months starting 96 CE.

Kiwi A nuclear fire-up {LANL}.

Some nuclear power is perfect for space. Radioisotope thermoelectric generators (RTGs) have been used in satellites and space probes since 1961. Mounting on extended arms reduces spacecraft radiation exposure and where cold dark space reduces solar cell effectiveness, RTGs get better thermal differentials. Select an isotope that has proper half-life for spacecraft life and the thermocouples just trickle out power at decay rate. Used by all outer solar system probes,

Tripping Over Engineering: Going Nuclear

they were also on an array of spacecraft due to particular benefits over solar cells. RTGs have been used terrestrially for remote arctic power. A few people even lived with these little nuclear generators in their chest cavities, giving early pacemakers a reasonable operating life compared to batteries of the time. [AMSE] has a collection of RTG mockups.

In 1965, SNAP-10A used thermoelectric elements through a NaK heat transfer medium operating as the only fission reactor in space. This had the potential for more output and control than an RTG. NASA is once again considering this type of reactor for future missions, some ideas never die. As an odd coincidence, on the side of the lot and not listed as a display item, [NMNSH] has an Atomics International SNAP-10A mockup in their restoration corner.

Nuclear with electric ion drive is already useful for long range spacecraft, but it doesn't quite equal the grandeur of a nuclear rocket plume breaking the surly bonds of earth. There is an old interest in nuclear thermal rockets for epic flights to places like Mars. In removing chemical reaction from the equation, it allows the propellant to be selected entirely for thermal and impulse characteristics without the annoying requirement of combustion management. This leads to hydrogen as a propellant with high heat transfer and high velocities out the nozzle reducing propellant mass needed. Compared to a nuclear jet with air as a working fluid, at least with hydrogen, there is no worry about high temperature oxidation in the reactor. That leaves the basic issues of a superheated hydrogen plume flaring in an air environment and the reentry/splashdown of a flaming radioactive reactor. A launch accident incurs an entirely new range of concerns.

Once you intend to use an area to detonate hundreds of nuclear weapons on the surface and underground, should someone ever say "Hey friend, you know where I could test some machinery that potentially sprays a lethal radioactive trail, just in its operation?", the only response is, "come on in, the water's fine". [NNSS] was a ready made location for testing. [NATM] has an array of models and artifacts.

Tripping Over Engineering: Going Nuclear

Phoebus 1B Nuclear Rocket
Nuclear Thermal Rocket Program
Tested 23 February 1967
Las Vegas NV [NATM]

Phoebus - AKA Apollo - Greek god who rides his chariot across the sky causing the rising and setting of the sun.

Phoebus 1B in transit, AEC/NASA photo.

In the NERVA program there was a stage of engine tests called Phoebus. Simply anchor down the reactor, pointing the nozzle straight up, add hydrogen and roll out the control drums.

Phoebus 1A operated for 10.5 minutes at 1,090 MW on 25 June 1965. Because the local radiation was so horrific, it affected the capacitance readings determining the hydrogen level on the tanks. While the loss of coolant burned out the core, it was found that the reactor was doing fine up to that point.

Phoebus 1B was designed for 1500 MW. It reached 1450 MW during a 30 minute test on 23 February 1967. During the process of the run, it only lost 1.5% of its fissile material out the nozzle. It does beg the question of how much fissile material is acceptable to blow out the nozzle. Now at [NATM].

Tripping Over Engineering: Going Nuclear

Nerva XE" Nuclear Rocket (AKA XE Prime)
Nuclear Thermal Rocket Program
Last of the nuclear rocket reactors
Huntsville AL [USSRC]

XE on reduced pressure simulator {Corliss 1971}.

Tripping Over Engineering: Going Nuclear

While Phoebus was part of the pure research effort, the next stage of NRX and Nerva reactors were technology demonstrators. XE" was the closest to a true operational reactor and considered to be a flight system configuration. Rather than firing up as previous, it fired down into a reduced pressure chamber simulating operating conditions expected of a space motor. Operated 4 December 1968 through 11 September 1969 with 24 starts over 40 runs. There were more test runs than starts due to cold flow tests for various conditions. The reactor was operated at full power of 1,140 MW for 3.5 minutes {Finseth 1991}. Engine #2 is on display, completed without operation as the program was canceled. Available if the Redstone Arsenal tour stops by Marshall Space Flight Center, bldg 4205.

Glossary

A-bomb
Atomic bomb, fission type, sometimes tritium boosted

ABM
Anti-Ballistic Missile. Originally with nuclear warheads, later as hit-to-kill.

AEC
Atomic Energy Commission. Keepers of the nuclear weapons program from 1947 to 1975. Nuclear weapons program remains under the Department of Energy.

CATOBAR
Catapult Assisted Take Off But Arrested Recovery. Typical modern aircraft carrier system used for non short/vertical take off and landing aircraft. Permits high aircraft performance with fewer concessions for low speed flight.

Critical Mass
Mass of fissile materials required to achieve prompt critical reaction. Critical mass varies with isotope and weapon design.

Criticality
Condition of self-sustaining fissile reaction.

DOE
Department Of Energy.

Fizzle
When a fission reaction reaches criticality with the energy blowing apart the critical mass before it reaches supercriticality. So, while there have been many criticality accidents, this is why a fission will typically behave like a reactor, but not like a warhead. This is also why Thin Man would not function as a bomb.

FOBS/MOBS
Fractional/Multiple Orbital Bombardment System. A Soviet system to alter traditional ballistic profile. Fractional would put the warhead into a partial low earth orbit for deorbit to target, Multiple would place it into a full orbit. These allowed a range and detectability avoidance beyond straight ballistic {Polmar 1975}. The Outer Space Treaty of 1967 declared orbital nuclear weapons illegal, the US declared that the FOBS was not in orbit.

H-bomb
Hydrogen bomb - fission/fusion - thermonuclear

KtE
Kilo (1,000) tonnage of TNT equivalent.

LABS
Low Altitude Bombing System - A toss bombing computer and method used to increase standoff from detonation. Typically referred to as Over the Shoulder (OTS) or more colloquially as the "Idiot's Loop".

Laydown Delivery
Weapon delivery with adequate parachute and reinforcement so that the weapon settles fully to the ground before detonation. Delayed laydown detonates after a time delay. The full ground contact is useful for bunker busting. It is also beneficial for providing increased escape time for the delivery aircraft, particularly with low speed/altitude delivery or high yield weapons.

MIRV
Multiple Independently targetable Reentry Vehicle. This permits multiple warheads each a level of guidance for more precision at impact. Not to be confused with MRV - Multiple Reentry Vehicle, effectively ballistic shotgun. Note that these are typically with a direct ballistic launch profile. While there are inherent accuracy capabilities within a direct ballistic path, see FOBS/MOBS.

MtE
Mega (1,000,000) tonnage of TNT equivalent

PAL
Permissive Action Links. Digital security integral to core systems on warheads which provide arming security. Often provided with a circuit shorting on fail or as a secondary safing through an external switch which fries the circuits. Some weapons also include a system which detects launch physics. Otherwise an armed weapon could hit its basic fuse conditions without an altitude drop, airspeed detection or time delay. While this may prevent immediate triggering, given time, tools and access, a competent technician could bypass the safety systems direct to the detonators.

Prompt Critical
Condition of generating prompt rather than decay neutrons at an exponential rate. Essential requirement of nuclear weapons.

Pu239
Plutonium 239 - Primary fissile isotope of plutonium, produced by transmutation of U238 while fissioning U235.

Pu240
Plutonium 240 - Fissile isotope of plutonium, produced with high energy reactor operations. Higher spontaneous criticality than Pu239 requiring use of implosion type weapons.

RATO
Rocket Assisted Take Off - The use of droppable rocket bottles for take off. Early nomenclature referred to as JATO.

SCRAM
Emergency fission shutdown of a nuclear reactor, typically with core insertion of boron or cadmium neutron absorbers as rods, balls or liquid solutions.

Strategic
Pertaining to activities not directly related to warfare engagement. Relates to action with manufacturing support capability and morale of civilian population.

Supercriticality
Nuclear fission at an exponentially increasing rate until stabilizing at high energy levels or self destructing.

Tactical
Pertaining to activities directly related to warfare engagement. Relates to action with military personnel, facilities and equipment.

tE
tonnage of TNT equivalent.

TNT Equivalent
Standardized warhead destructive power. The energy of a nuclear blast as compared to 100% TNT (Trinitrotoluene). The initial Trinity test was preceded by a 100 ton TNT test to calibrate equipment for the nuclear detonation. It provides a standard of reference for high energy events, such as industrial explosions and meteor strikes. It should be noted that it has limits due to other characteristics of nuclear weapons. Besides the direct blast, extensive damage can arise from EMP, direct radiation, thermal flash, incendiary and fallout.

Two Man Rule
Nuclear security requiring two crew in close contact with each other while in secure areas. Also known as "No Lone Zone".

U235
Uranium 235 - Only natural fissile isotope of uranium. Exists at 0.711% of natural uranium.

U238
Uranium 238 - Non-fissile isotope of uranium. Exists at 99% of natural uranium. Bred to generate Pu239.

Yield
Explosive power of a detonation, see TNT Equivalent.

Tripping Over Engineering: Going Nuclear

[Location by State]

AL	Huntsville	[USSRC]
AZ	Tucson	[PASM]
AZ	Tucson	[TMM]
CA	Merced	[CAM]
CA	Ridgecrest	[USNMAT]
CA	San Francisco	[NMS]
CT	Groton	[SFM]
DC	Dulles	[NASMuh]
DC	Navy Yard	[NMUSN]
FL	Eglin AFB	[AFAM]
FL	Pensacola	[NNAM]
GA	Robins AFB	[MOA]
ID	Idaho Falls	[INL]
IL	Chicago	[MSI]
MT	Great Falls	[GTF]
NC	Charlotte	[CaAM]
NE	Ashland	[SACSM]
NM	Alamogordo	[NMMSH]
NM	Albuquerque	[NMNSH]
NM	Los Alamos	[BSM]
NM	Las Cruces	[WSMR]
NV	Las Vegas	[NATM]
NV	Las Vegas	[NNSS]
NY	NYC	[ISASM]
OH	Dayton	[NMUSAF]
OR	McMinnville	[EASM]
SD	Ellsworth AFB	[SDASM]
SD	Philip	[MMNHS]
TN	Oak Ridge	[AMSE]
WA	Richland	[HS]

Tripping Over Engineering: Going Nuclear

[Location by Site]

Some of these are on military or secure facilities. They may have a variety access and screening, some are extremely restrictive in both schedule and security. These may also lock down for a variety of reasons. GPS location and routing may not correlate to the security access. Some particularly large sites may have tours with a range of variability in the stops. Check before you go and be prepared for an occasional delay.

For weapons and aircraft, survivors are often listed on Wiki or Veteran's unit websites. If you are looking for something specific nearby, this gives a good start. Also note that artifacts may be removed for restoration, relocated to another museum or in some cases scrapped. The museum website may not be in alignment with current inventory. If you are traveling out of your way for a particular item, I recommend confirming its location with staff. Military combat facilities like ships and missile sites were originally designed for fit and healthy young men, confirm accessibility if you have concerns.

Many of these locations have docents with specific relationships to the facility. Do not hesitate to ask any questions. They may provide a particular insight into the daily operations. And that enthusiasm may at times lead to access to portions of the site that they do not typically allow tourists.

[AFAM]
Air Force Armament Museum: Eglin AFB, FL
Visited April 2018. Outside of base, no particular restrictions. Excellent aircraft weapons collection. Includes items that some museums may have avoided as unpopular. Near [NNAM].
http://www.afarmamentmuseum.com/

[AMSE]
American Museum of Science & Energy: Oak Ridge, TN
Visited March 2017. Excellent Oak Ridge artifacts. Provides daily tour of Oak Ridge National Lab.
http://amse.org/

Tripping Over Engineering: Going Nuclear

[BSM]
Bradbury Science Museum: Los Alamos, NM
Visited September 2012 and October 2018.   Smaller museum with unique collection.   There are currently no tours inside LANL.   NPS is in discussions with LANL on the possibility of touring NPS listed historic sites.
http://www.lanl.gov/museum/
http://www.lanl.gov/museum/public.php
http://www.nps.gov/mapr/manhattan-project-los-alamos.htm

[CaAM]
Carolinas Air Museum: Charlotte, NC
Have not visited, but displays some particularly unique items.
http://www.carolinasaviation.org/

[CAM]
Castle Air Museum: Atwater, CA
Have not visited, but has extensive collection.
http://www.castleairmuseum.org

[EASM]
Evergreen Aviation & Space Museum: McMinnville OR
Visited August 2013.   Amazing collection.   If they still offer the Spruce Goose cockpit tour, take it!
http://evergreenmuseum.org

[GTF]
Great Falls International Airport: Great Falls, MT
Have not visited.   It has the nuclear test F-89J on display with several other aircraft.   The terminal supposedly has the world's largest collection of model planes.

[HS]
Hanford Site: Richland, WA
Clean up site tours with limited registration and restricted access.
http://www.hanford.gov
Hanford B Reactor
Visited August 2013. Stand in front of the reactor, it is a sublime experience. B Reactor tours have limited registration and restricted access. T Plant (Queen Marys) had a 2013 anniversary tour, hopefully there will be more in the future.
http://manhattanprojectbreactor.hanford.gov
http://b-reactor.org/

[INL]
Idaho National Laboratory: Idaho Falls ID
Visited August 2013. EBR-1 and HTRE reactors up close and personal. No charge, regular hours, outside of lab security.
http://www.inl.gov/

[ISASM]
Intrepid Sea, Air & Space Museum: New York NY
Visited 2005. Special collection.
http://intrepidmuseum.org

[MMNHS]
Minuteman Missile National Historic Site: Philip, SD
Visited June 2011. Must register at museum for excellent tour of the launch control bunker with silo nearby.
http://www.nps.gov/mimi/index.htm

[MOA]
Museum of Aviation: Robins AFB, GA
Visited April 2018. Excellent collection with possession of some notable historic aircraft.
http://www.museumofaviation.org/

[MSI]
Museum of Science and Industry: Chicago, IL
Visited 2005. The MSI and Shedd Aquarium are amazing at so many levels, must see.
http://www.msichicago.org

Tripping Over Engineering: Going Nuclear

[NASMuh]
National Air and Space Museum, Udvar Hazy Center: Chantilly, VA
Visited 2004. Adjacent to Washington Dulles International Airport. Excellent collection like all Smithsonian assets. Has some very specific historic aircraft. While in DC, also stop by the original Air and Space Museum.
http://airandspace.si.edu/udvar-hazy-center

[NATM]
National Atomic Testing Museum: Las Vegas, NV
Visited 2009. Small, but unique museum.
http://nationalatomictestingmuseum.org/

[NMMSH]
New Mexico Museum of Space History: Alamogordo, NM
Visited October 2018. Unique rocket, guidance and space research collection.
http://www.nmspacemuseum.org/

[NMNSH]
The National Museum of Nuclear Science & History: Albuquerque, NM
Visited September 2010 and October 2018. Excellent collection.
http://www.nuclearmuseum.org/

[NMS]
Nike Missile Site: Golden Gate National Recreation Area, CA
Visited December 2009. One of a kind site, at the time riding the elevator with the missile down into the hangar.
http://www.nps.gov/goga/nike-missile-site.htm

[NMUSAF]
National Museum of the USAF: Wright-Patterson AFB, OH
Visited 2005. The collection is now outside of base security containing historic aircraft and singular prototypes.
http://www.nationalmuseum.af.mil/

[NMUSN]
National Museum of the US Navy: Navy Yard, DC
I have not visited. It apparently requires specific base pass for entry.
http://history.navy.mil/content/history/museums/nmusn.html

[NNAM]
National Naval Aviation Museum: Pensacola, FL
Visited 1988, April 2018 & sometime in between. Looking forward to the next visit. Inside Air Station Pensacola with entrance and exit only through West Gate and identification required for entry. This end of base has Museum, Fort Barrancas and Lighthouse. Near [AFAM].
http://www.navalaviationmuseum.org/

[NNSS]
Nevada National Security Site: Las Vegas, NV
Formerly, Nevada Test Site
Have not visited, but scheduled for February 2019.
Restricted high security area. Tours fill up quickly, require advance registration with specific security and access requirements.
http://www.nnss.gov/pages/PublicAffairsOutreach/NNSStours.html

[PASM]
Pima Air & Space Museum: Tucson, AZ
Davis Monthan AFB tours through PASM.
Visited September 2012 and November 2015. Pima has great collection including limited prototypes. Monthan is always something new depending on the aircraft being decommissioned.
http://www.pimaair.org/

[SACSM]
Strategic Air Command & Aerospace Museum: Ashland, NE
I have not visited, but they appear to have a good collection with some rarer well documented aircraft.
http://sacmuseum.org

Tripping Over Engineering: Going Nuclear

[SDASM]
South Dakota Air and Space Museum: Ellsworth AFB, SD
Visited June 2011. Museum is outside of base. Special security screening for Ellsworth AFB and Minuteman II training silo. Tour is first come first served. Silo is worth the trip.
http://www.sdairandspacemuseum.com/

[SFM]
Submarine Force Museum: Groton, CT
Visited 2006. Home to the USS Nautilus and a great collection of submarine specific displays.
http://ussnautilus.org/

[TMM]
Titan Missile Museum: Sahuarita, AZ
Visited September 2012 and November 2015. Second time took the top to bottom tour. 54 Titan II sites built and the only one remaining intact with systems and missile. Personally recommend taking the most extensive tour available.
http://www.titanmissilemuseum.org/home

[USNMAT]
US Naval Museum of Armament and Technology: Ridgecrest, CA
Have not visited. Museum has moved off base so pass not required. A collection of naval weapons testing back to 1943.
http://chinalakemuseum.org

[USSRC]
U.S. Space & Rocket Center: Huntsville, AL
Visited March 2017 and April 2018. Stand alone museum outside of Redstone Arsenal with charge for museum and tours. The Saturn V alone is worth a trip. Tours of Redstone Arsenal/Marshall Space Flight Center require tickets and identification. Note that while the Redstone tour typically visits the Redstone test stands and the Space Flight Center, other stops may vary.
http://www.rocketcenter.com/

[WSMR]
White Sands Missile Range, NM
Visited October 2018. Museum on base with security clearance required and regular hours. Extensive collection of weapons and test rockets.
http://wsmr-history.org

Trinity Site Tour
Visited October 2012 and October 2018. Currently open two days per year, so plan accordingly. A surreal experience. Note also that [NMNSH] has tours from Albuquerque.
http://www.wsmr.army.mil/Trinity/Pages/Home.aspx

Very Large Array (VLA)
Visited October 2012 and October 2018. VLA radio telescope has a partner open house on the same tour days as Trinity with more extensive than typical tours. I recommend Trinity when they open the gate and VLA in the afternoon.
http://public.nrao.edu/visit/very-large-array/#tours

{Sources}

This did not start as a research document, but the farther I went down the rabbit hole, variances would sometimes occur between well pedigreed sources. This lead to more intensive citing as the writing progressed. I have tried to reference as specifically as possible. Many of the sources are excellent additional reading.

I endeavored to find appropriate historic photographs where possible. All graphics are referenced for source.

Wikipedia may have issues with information stability in more subjective material, but in this area I found it to have well referenced information. I used Wiki extensively as a launch point to source documents.

{AFHRA}
Air Force Historical Research Agency. Historic photo collection. These are public domain.
Enola Gay:080128-F-3927S-065
http://www.afhra.af.mil/Photos/igphoto/

{AMC 1976}
U.S. Army Material Command, National Technical Information Service (1976 January). Engineering Design Handbook: Recoilless Rifle Weapon Systems, AMC Pamphlet AMCP 706-238. Para.1-5.2.
http://www.dtic.mil/dtic/tr/fulltext/u2/a023513.pdf

{ANL}
Argonne National Laboratory. ANL defers photo storage to flickr. These are public domain.
EBR-1 start
http://www.anl.gov/
http://www.flickr.com/photos/argonne/

{Baggott 2010}
Baggott,J. (2010) The First War of Physics: The Secret History of the Atomic Bomb 1939-1949, New York, NY: Pegasus Books. ISBN 978-1-60598-084-3   Supporting reference not cited.

{BI}
Brookings Institution.   These are public domain.
W48 credit DOE
http://www.brookings.edu/u-s-nuclear-weapons-photo-gallery/

{Bragg 1961}
Bragg,J. (April 1961) Development of the Corporal: The Embryo of the Army Missile Program Volume I, Historical Monograph No.4, Army Ballistic Missile Agency, Redstone Arsenal.
http://www.dtic.mil/dtic/tr/fulltext/u2/a586733.pdf

{Bullard 1965}
Bullard,J. (15 October 1965) History of the Redstone Missile System, AMC 23 M, Army Missile Command, Redstone Arsenal.   Supporting reference not cited.
http://www.dtic.mil/dtic/tr/fulltext/u2/a434109.pdf

{Corliss 1971}
Corliss,W. & Schwenk,F. (1971) Nuclear Propulsion for Space, An Understanding the Atom Series Booklet, US Atomic Energy Commission.   LOC 79-171030 1968;1971(rev.)
http://www.osti.gov/includes/opennet/includes/Understanding%20the%20Atom/Nuclear%20Propulsion%20for%20Space%20V.2.pdf.

{Dobbs 2008}
Dobbs,M. (2008) One Minute to Midnight, New York, NY: Knopf Random House. ISBN 978-1-4000-4358-3   pp. 10, 14, 31, 39, 58, 249, 251, 256, 276, 281.   Excellent details on the Cuban Missile Crisis.

{DOE}
U.S. Department of Energy, Office of History and Heritage Resources. DOE defers photo storage to flickr. These are public domain.
CP-1:HD.5A.038
X-10:5221-2
Trinity:TR00224
Y-12:HD.30.844
Hanford:HD.4A.130
http://www.energy.gov/management/historical-photographs
http://www.flickr.com/photos/departmentofenergy/

{DOE 1943}
U.S. Department of Energy (1943 September). "Memorandum Covering Technical Basis for Work under Contract Numbered W-7412 eng-1 between United States of America and E.I. du Pont de Nemours & Company", DOE Document No. HAN-43508.

{DOE 2006}
U.S. Department of Energy, Office of History and Heritage Resources (2006 September). Battlefield of the Cold War, The Nevada Test Site Volume I, Atmospheric Nuclear Weapons Testing: 1951-1963, DOE Document No. DOE/MA-0003. pp. 30, 71, 99, 100.
http://energy.gov/sites/prod/files/DOENTSAtmospheric.pdf

{DOE 2015}
U.S. Department of Energy, National Nuclear Security Administration Nevada Field Office (2015 September). United States Nuclear Tests: July 1945 through September 1992, DOE Document No. DOE/NV-209-REV 16. p.xv.
http://nnss.gov/docs/docs_LibraryPublications/DOE_NV-209_Rev16.pdf

{DTIC}
Defense Technical Information Center.
http://www.dtic.mil/dtic/search/advanced_search.html
Palomares Summary Report 15 January 1975, Field Command Defense Nuclear Agency, ADA955702, p.104 fig 3-19.
http://www.dtic.mil/dtic/tr/fulltext/u2/a955702.pdf

{DTRA}
U.S. Defense Threat Reduction Agency. DTRA defers photo storage to flickr. These are public domain.
Buster Jangle Easy:11051951
http://www.dtra.mil/
http://www.flickr.com/photos/dod_dtra/

{Feder 2017}
Feder,K. (2017) Ancient America: fifty archaeological sites to see for yourself, Lanham MD: Rowman & Littlefield. ISBN 9781442263123
LCCN 2016038670
Great information, try to make some of these stops while you are between my sites.

{Finseth 1991}
Finseth,J. (February 1991) Rover Nuclear Rocket Engine Program: Overview of Rover Engine Tests: Final Report, Contract NAS 8-37814, File No. 313-002-91-059, Marshal Space Flight Center. pp.63,112.
https://ntrs.nasa.gov/archive/nasa/casi.ntrs.nasa.gov/19920005899.pdf

{Friedman 1983}
Friedman,N. (1983) U.S. Aircraft Carriers, Annapolis MD: United States Naval Institute. ISBN 0-87021-739-9 pp 244-247.

{Gibson 1996}
Gibson,J. (1996) Nuclear Weapons of the United States: An Illustrated History, Atglen PA: Schiffer Publishing Ltd. ISBN 0-7643-0063-6 LCCN 96-67282 p.14,111.

{Grier 2012}
Grier,P. (2012, January). The B-52 Gunners, Air Force Magazine.
http://airforcemag.com/MagazineArchive/Pages/2012/January%202012/0112gunners.aspx

{Gulliver 2011}
Gulliver,V. (2011), "The Historic Flight of the Truculent Turtle", Retrieved 17 June 2018. This was a PDF document which stated originally written for Patrol Squadron Two website and printed in condensed form in the "Wings of Gold" magazine Spring 2011.
http://maritimepatrolassociation.org/documents/heritage/Truculent_Turtle_1946.pdf

{Hansen}
Hansen,C (Unknown), U.S. Nuclear Weapons: The Secret History. This is material that was directly attributed to the work of Chuck Hansen in other documents. I was unable to access this book. Therefore, I can not confirm the accuracy of secondary to the primary source. Mr. Hansen accumulated the largest collection of nuclear weapons information through the Freedom of Information Act. This collection is archived in its entirety in the National Security Collection of George Washington University, Washington DC.

{Heinlein 1959}
Heinlein,R (1959), Starship Troopers, New York NY: Penguin Random House. Politically questioned when it was released and now mandatory reading at West Point. Good science fiction is always a greater commentary of its present than a predictor of the future.

{Herken 1990}
Herken,G. (1990, April/May), "The Flying Crowbar", Air & Space Magazine, Retrieved 7 Aug 2018, p.28.
http://www.merkle.com/pluto/pluto.html

{Higham 1975}
Higham,R. & Siddall,A. (Eds.) (1975). Flying Combat Aircraft of the USAAF-USAF, Ames, IA: The Iowa State University Press. ISBN 0-8138-0325-X   F-89 & B-58.

{Keller 1960}
Keller,M. (1960, July). Little John - "THE MIGHTY MITE", Artillery Trends, 20-25.
http://sill-www.army.mil/firesbulletin/archives/1960/JUL_1960/JUL_1960_FULL_EDITION.pdf

{Kubrick 1964}
Kubrick,S. Dir. (1964), "Dr. Strangelove or: How I Learned to Stop Worrying and Love the Bomb", Columbia Pictures, Film. On the subject of the Cold War fears, this is probably the most must see movie.

{LANL}
Los Alamos National Lab. Historic photos.
Public domain:
Fermiac
http://permalink.lanl.gov/object/tr?what=info:lanl-repo/lareport/LA-UR-00-2532
Creative Commons-NonCommerial-NoDerivatives:
8" Atomic Shell
Upshot Knothole Grable
Kiwi A
http://www.flickr.com/photos/losalamosnatlab/

{LB}
Lindsey Bredemeyer, all rights reserved. In lieu of finding the historic photo I want, you have to look at my camera work.

{LOC}
Library of Congress. Historic photo collection, significant ones on line.   These are public domain.
Redstone: HAER ALA,45-HUVI.V,7A--40
http://www.loc.gov/pictures/search/

{Lodge 1945}
Lodge,J. (1945, July), Biggest Gun on Wheels, Popular Science, pp.76-79.

{Lonnquest 1996}
Lonnquest,J. & Winkler,D. (November 1996) To Defend and Deter: The Legacy of the United States Cold War Missile Program, USACERL Special Report 97/01, Department of Defense Legacy Resource Management Program. pp.211,419.
http://www.dtic.mil/dtic/tr/fulltext/u2/a337549.pdf

{Mahaffey 2014}
Mahaffey,J. (2014) Atomic Accidents: A History of Nuclear Meltdowns and Disasters: From the Ozark Mountains to Fukushima, New York, NY: Pegasus Books. ISBN 978-1-60598-492-6 p.300 Excellent source on reactor, processing and weapons accidents.

{Michel 2003}
Michel,M. (2003, May), "Exit Strategy", Air & Space Magazine, Retrieved 16 June 2018.
http://www.airspacemag.com/military-aviation/exit-strategy-4410379/?all

{Miller 2001}
Miller,J. (2001) Nuclear Weapons and Aircraft Carriers: How the Bomb Saved Naval Aviation, Washington DC: Smithsonian Institution. ISBN 1-56098-944-0 pp.37,58-67,80-87,94-98,108,126-129. Excellent details.

{NAC}
National Archives Catalog. These are public domain.
NS Savannah: NAC; 542141 Local; 326-NS-56
USS Nautilus: NAC; 521003
A3D: NAC; 6483221 Local; 330-CFD-DN-SC-93-02296
Little Boy: NAC; 519394 Local; 77-BT-115
http://www.archives.gov/
http://catalog.archives.gov/

{NASA}
National Aeronautics and Space Administration. Historic photo collection. These are public domain.
XB-70: 359356main_ECN-792
http://www.nasa.gov/centers/armstrong/multimedia/imagegallery/index.html

Tripping Over Engineering: Going Nuclear

{NHHC}
Naval History and Heritage Command. Historic photo collection.   These are shown as US government photographs for US government purposes, so public domain.
Fat man: 80-G-396229
AJ: NH 97459
AD: NH 94701
A-5: USN 1130917
P-2V: 80-G-707207
Turtle: 80-G-703094
TDR-1: USN 1053775
LOON: NH 72680
Regulus: NH 72679
BB-63: 80-G-K-12603
http://www.history.navy.mil/
http://www.history.navy.mil/our-collections/photography.html

{NNAM}
National Naval Aviation Museum. Historic photo collection, significant ones on line.   These are public domain.
Enterprise: NNAM.1996.488.125.008
http://collections.naval.aviation.museum/emuwebdoncoms/pages/doncoms/AdvQuery.php

{NMUSA}
National Museum of the US Army. Historic articles and photos. These are public domain.
Davy Crockett
http://armyhistory.org/

{NMUSAF}
National Museum of the US Air Force. Historic photo collection, significant ones on line. These are public domain. VIRIN numbers shown.
Lucky Lady II: 061215-F-1234S-002
C-99: 120409-F-XN622-001
Titan II: 140124-F-DW547-006
AGM-28: 061127-F-1234S-025
SM-65: 050406-F-1234P-014
B-47: 061024-F-1234S-011
B-58: 061101-F-1234P-016
B-36: 120409-F-XN622-002
B-52: 061127-F-1234S-003
B-25: 050607-F-1234P-018
XF-84: 080226-F-1234S-004
SM-62: 061218-F-1234P-003
CIM-10: 090603-F-1234P-002
YF-89A: 060829-F-1234S-044
OTS: 050419-F-1234P-023
LGM-30: 090108-F-1234P-009
Crew: 090108-F-1234P-010
http://www.nationalmuseum.af.mil/Upcoming/Photos/

{Polmar 1975}
Polmar,N. (1975) Strategic weapons: An Introduction, New York NY: Crane, Russak & Company, Inc. ISBN 0-8448-0822-9   pp. 12, 14, 28, 54, 90, 108.

{Polmar 2004}
Polmar,N. (2004) Cold War Submarines, Dulles VA: Brassey's, Inc. ISBN 1-57488-594-4   pp. 58, 60, 61, 86, 87, 103, 130, 131.

{Polmar 2009}
Polmar,N. & Norris,R. (2009) The U.S. Nuclear Arsenal: A History of Weapons and Delivery Systems since 1945, Annapolis MD: Naval Institute Press. ISBN 978-1-55750-681-8   pp. 36, 37, 42, 47, 75, 76, 77, 84, 96, 99, 166, 185, 229.

{RA}
US Army Redstone Arsenal. Historic photo collection, significant ones on line.   These are public domain.
http://history.redstone.army.mil/redstone/redstone_1st_tac_firing_6jul61_01.jpg
http://history.redstone.army.mil/miss/honest_john_05.jpg
http://history.redstone.army.mil/miss/nike/hercphotos/nike_herc_40.jpg
http://history.redstone.army.mil/miss/corporal_1958_wsmr_01.jpg

{Ramsbotham 1989}
Ramsbotham,O. (1989) Modernizing NATO's Nuclear Weapons: 'No decisions have been made', Hampshire UK: The Macmillan Press LTD.   ISBN 978-0-333-49672-5   pp.60-63.

{Rausa 1982}
Rausa,R. (1982) Skyraider, The Douglas A-1: Flying Dump Truck, Baltimore MD: Nautical & Aviation Publishing Company of America.   ISBN 0-933852-31-2

{Rhodes 1986}
Rhodes,R. (1986), The Making of the Atomic Bomb, New York NY: Touchstone, Simon Schuster.   ISBN 0-671-44133-7   pp. 11, 92, 340, 373, 420, 433, 436, 466, 556, 558, 600, 625, 631, 656, 681.

{Rhodes 2004}
Rhodes,J. (2014, Q3), "C-130 Willis Hawkins and the Genesis of the Hercules", Code One Magazine, 19(3), pp.16-21.

{Rotter 2008}
Rotter,A. (2008) Hiroshima: The World's Bomb, New York, NY: Oxford University Press.   ISBN 978-0-19-280437-2   pp. 147, 175, 188, 208, 210, 743, 745.

{Schlosser 2013}
Schlosser,E. (2013) Command and Control: Nuclear Weapons, the Damascus Accident, and the Illusion of Safety, New York, NY: The Penguin Press. ISBN 978-1-59420-227-8   pp. 5, 23, 28, 32, 41, 46, 50, 65, 83, 88, 96, 97, 99, 130, 140, 148, 150.   Covers the Little Rock Titan II explosion and covers a large range of nuclear development and policy. PBS produced and excellent American Experience episode also called "Command and Control".

{Schwiebert 1965}
Schwiebert,E. (1965) A History of the U.S. Air Force Ballistic Missiles, New York NY: Frederick Praeger Publishers.   LCCN 65-14058   pp. 77, 153, 193, 217.

{Serchak 1980}
Serchak,W. (1980, March-April). "Artillery Fired Atomic Projectiles - A Field Artilleryman's Viewpoint", Field Artillery Journal, 7-8.
http://sill-www.army.mil/firesbulletin/archives/1980/MAR_APR_1980/MAR_APR_1980_FULL_EDITION.pdf

{Silverstone 2013}
Silverstone,P. (2009) The Navy of the Nuclear Age 1947-2007: US Navy Warship Series, New York, NY: Routledge. ISBN10 0-415-97899-6   pp.xxxii,xxxiii.

{Smith 1998}
Smith,R. (1998) Seventy-five years of Inflight Refueling: Highlights, 1923-1998, Air Force History and Museums Program.   ISBN 0-16-049779-5   p.44.
http://www.amc.af.mil/Portals/12/documents/AFD-141230-027.pdf

{Sublette}
Sublette,C. (), Section 8.0 The First Nuclear Weapons, Nuclearweaponarchive.org, Retrieved 14 July 2018.
http://nuclearweaponarchive.org/Nwfaq/Nfaq8.html
http://nuclearweaponarchive.org/Usa/Tests/

{Thornton 1962}
Thornton,G. & Rothstein,A. (June 1962) Comprehensive Technical Report, General Electric Direct-Air-Cycle, Aircraft Nuclear Propusion Program, Program Summary and References, APEX-901 General Electric for AEC.
http://leehite.org/anp/documents.htm

{Thornton 1963}
Thornton,G. (February 1963) Introduction to Nuclear Propulsion, Lecture 1 - Introduction and Background, General Electric for MSFC NASA.   p.29.
http://ntrs.nasa.gov/archive/nasa/casi.ntrs.nasa.gov/19640019868.pdf

{Tirpak 1994}
Tirpak,J. (1994, April). "The Secret Squirrels", Air Force Magazine.
http://www.airforcemag.com/MagazineArchive/Pages/1994/April%201994/0494squirrels.aspx

{USNI 2011}
Blog (2011 February 6), "Aviation Centennial: Neptune's Atomic Trident (1950)", US Naval Institute, Retrieved 18 June 2018.
http://blog.usni.org/posts/2011/02/06/naval-aviation-centennial-neptunes-atomic-trident-1950

{USPTO}
US Patent and Trademark Office.
Spark Gap: 3,956,658
Bridgewire: 3,040,660
http://www.uspto.gov/

{WIKI}
Wikipedia online, photos noted on Wiki as public domain. Supersonic Low Altitude Missile - noted as from NNSS tour booklet no longer linked.
http://en.wikipedia.org/wiki/Main_Page

www.ingramcontent.com/pod-product-compliance
Lightning Source LLC
Chambersburg PA
CBHW031618210526
45464CB00004B/1628